200010005087 28

D0719868

GAYLORD

ALSO BY DAVID BODANIS

THE SECRET GARDEN

WEB OF WORDS

THE SECRET HOUSE

THE BODY BOOK

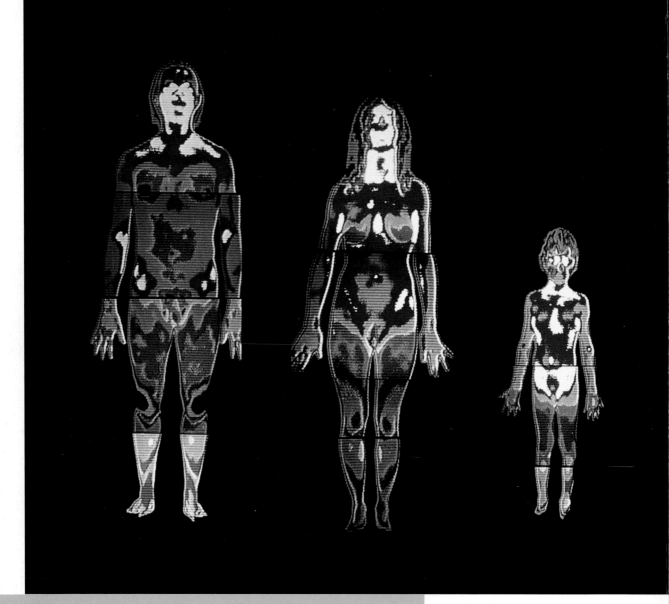

david bodanis

simon & schuster

612
Bod

KOREMATSU CAMPUS

the secret family

twenty-four hours inside the

mysterious world of our

minds and bodies

SIMON & SCHUSTER

Rockefeller Center

1230 Avenue of the Americas

New York, NY 10020

Copyright © 1997 by David Bodanis

All rights reserved, including the right of reproduction in whole or in part

in any form.

SIMON & SCHUSTER and colophon are registered trademarks of Simon & Schuster Inc.

Designed by Karolina Harris

Manufactured in the United States of America

10   9   8   7   6   5   4   3   2   1

Library of Congress Cataloging-in-Publication Data

Bodanis, David.

   The secret family: twenty-four hours inside the mysterious world of our minds and

bodies / David Bodanis.

     p.   cm.

   Includes index.

   1. Human biology—Popular works.   2. Human ecology—Popular works.

I. Title.

QP38.B594   1997   97-14809   CIP

612—dc21

ISBN 0-684-81019-0

In Blessed Memory, H. Samuel Bodanis (1901–1966)

In Bountiful Joy,
Samuel Adam Bodanis, born 1993
Sophie Beatrice Bodanis, born 1996

contents

introduction
and acknowledgments

I got the idea for this book, or at least the first hints of it, years and years ago
when I was a little boy in Chicago. I was the youngest of a big family (five
older sisters) and delighted at being surrounded by these great adult crea-
tures, who breezed in with boyfriends or homework, with their shopping, and
phone calls, and mysteriously wonderful school parties and dates to arrange.
I also marveled at the way our family was a machine that somehow—through
a mix of discussions, arguments, and agreements—managed to supply all the
living requirements of its members: the blankets and sweaters, the words,
food, shoes, books, milk deliveries, and all the rest.

By the end of each summer, though, we were bursting out of our house
and often our previous year's friendships, and it was time to leave for the
long drive to my grandmother's farm, packing in the suntan lotions, sand-
wiches, slacks and hair bands, boyfriends' phone numbers, writing paper,
iced orange juice, books to read, and caramels, along with the six children
and two parents. (Pets, though appreciated at home, were not allowed to
ride.) At first I couldn't understand why navigation involved having a map
spread out in the front seat: surely it would be simplest for the government to
put up signs saying TO GRANDMA'S FARM at the appropriate highway inter-
sections. I was even willing to have a few extra signs added, including refer-
ences to my best school friends' families, so that they would be able to
navigate too.

And then—was I eight?—it suddenly hit me that it wasn't just my fam-
ily in a car, or my neighborhood's families, but thousands, millions, of Ameri-
cans driving here and there on the highways, and that each one was part of a

family that we would most likely never meet, but that still had to face almost all the supply and assembly and sorting operations my own family did.

It was a staggering thought. Were there really other little boys with crew cuts and white T-shirts who had big sisters that teased them? There would have to be an unbelievable number of white T-shirts produced in America to dress them all. And then, where did all those other sisters live? How could they function properly without me around? I tried to imagine how many newspapers and gym shoes and cereal bowls would have to be made to supply all the families: how many school principals, and milk delivery men, and teachers of the ethical principles it was so important to learn.

Time passed, and I continued to be fascinated by families. As I studied a little science I delighted in the image of humans making do not alone, but in family groupings, as we live surrounded by invisible chemicals, energies, and microbes. Later I worked in different countries, and noticed how varying families would cope with the great crashing waves of emotions or financial need or politics that slammed up against them, how events sluiced through in totally different fashions, depending on how the family was constructed. And then—nicest of all—I started my own family.

And then I knew it was time to write.

Most of the research for this book was conducted at the University of London's Senate House Library, the University of London Science Library, and Oxford's Radcliffe Science Library, though I also was helped by library staff and resources at the Science Reference Division of the British Library, as well as at University College London, the London School of Hygiene and Tropical Medicine, Imperial College, the London Business School, the Institute of Education at the University of London, London's Science Museum, and the Wellcome Trust.

In almost every area there were a great number of professional journals to help, and my PowerBook merrily beeped warning tunes from Nina Simone, to the delight and occasional astonishment of my fellow researchers, as the hours taking notes wore down its battery. The most useful journals and their generally unpaid review staffs deserve a special thanks, including: *Acarologia; Advances in Consumer Research; American Demographics; American Journal of Clinical Nutrition; American Journal of Pediatrics; American Psychologist; American Scientist; Annals of the New York Academy of Science; Annals of Occupational Hygiene; Annual Review of Microbiology; Annual Review of Nutrition; Annual Review of Psychology; Appetite, Archives of Environmental Health; Behavior Genetics; Behavioral Pharmacology; British Journal of Dermatology; British Journal of Social and Clinical Psychology; British Journal of Social Psychology; Building Services Journal; Bulletin of the American Meteorological Society; Bulletin of the Psychonomic Society; Chemical Senses; Clinical Allergy; Developmental Psychology; Ecological Sociobiology; Environmental Behavior; Environmental Entomology; Environmental Health Perspectives; Ethnology; Ethnological Sociobiology; Gut; Human Neurobiology; Human Relations; Indoor Environment; International Archives of Occupational and Environmental Health; International Biodeterioration; International Journal of Psychology; Journal of Advertising; Journal of Chemical Ecology; Journal of Community and Applied Social Psychology; Journal of Consumer Research; Journal of Experimental Social Psychology; Journal of Family Psychology; Journal of Genetic Psychology; Journal of Marketing Research; Journal of Marriage and the Family; Journal of Nonverbal Behavior; Journal of Nutritional Medicine; Journal of Occupational Psychology; Journal of Pediatrics; Journal of Personality and Social Psychology; Journal of Psychiatric Research; Journal of Psychology; Journal of the American Medical Association; Metabolism; Mycological Research; Nature; Neuroscience Biobehavior Reviews; New Scientist; Nutrition Research; Oecologia; Pediatrics; Personality and Social Psychology Bulletin; Proceedings of the Society of Experimental and Biological Medicine; Psychological Bulletin; Psychophysiology; Science; Scientific American; Semiotica; Sensory Processes; Social Psychology Quarterly; Transactions of the British Mycological Society;* and *Trends in Pharmacological Sciences.*

Not everything could be found in printed sources, and many people went out of their way to be helpful as my questions multiplied. I liked finding that Luther had worried about the imminent decline of marriage, 400 years before our current Op-Ed pieces about it; now I enjoyed finding out such information as what had been in those massive quantities of gum I'd chewed as a kid. Disgruntled chemists and engineers at several major consumer products companies were an especially useful source for a range of matters, though thanking them by name would not aid their professional advancement. Individuals and organizations who can be more easily thanked include: Dr. Peter Addyman, York Archaeological Trust; Professor R. M. Alexander, the authority on biomechanical optimization, Department of Pure and Applied Zoology, University of Leeds; Sue Cavill at the British Psychological Association; British Telecom; Dr. Ian Burgess, Medical Entomology Centre, Cambridge University; Professor B. A. Bridges, Medical Research Council Cell Mutation Unit, University of Sussex; Professor Bradbury, Department of Physiology, King's College London, for details on blood-brain interactions; Dr. J. Empson, Department of Psychology, University of Hull; Jan Pieter Emans, Ciba Resources Center, London; Dr. R. C. Garner, The Jack Birch Unit for Environmental Carcinogenesis, Department of Biology, University of York; Dr. Malcolm Green, Royal Brompton Hospital; Adrianne Hardman, Department of Physical Education, Sports Science and Recreation Management, Loughborough University; Judy Hildebrand, Institute of Family Therapy, London; Professor Stephen Holgate, Immunopharmacology Group, Southampton General Hospital; Dr. Lewis Smith of the MRC Toxicology Unit and Dr. Paul Harrison at the University of Leicester; Professor Andre McLean, Toxicology Laboratory, Department of Clinical Pharmacology, University College, London; Dr. Gene Mooney, London Foot Hospital; the University of Munich's Institute for Empirical Pedagogy and Psychology; Oxford Brookes University's Centre for Science of Food and Nutrition; Dr. Peter Rogers, Institute of Food Research, Reading; The Royal Mail; Professor Mick Rugg, cognitive development, St. Andrews University; Dr. John Sloboda, music and auditory comprehension, Keele University; Dr. Peter Stratton, Leeds Family Therapy and Research Centre, University of Leeds; Dr. John Timbrell, School of Pharmacy, University of London; Dr. Richard Wood, Imperial Cancer Research Fund, Clare Hall Labs; Professor Colin Walker, Department of Biochemistry and Physiology, University of White Knights, Reading; Dr. Saffron Whitehead, St. George's Hospital Medical School. Special thanks go to Tim Lobstein and Sue Dibb at the London Food Commission. When Her Majesty's Government insisted there was no possible danger from Bovine Spongiform Encephalitis—"Mad Cow Disease"—oh, did the

Commission have information proving the contrary, as well as leads for almost every other food topic conceivable. The Islington-based Women's Environmental Network, and Ann Link in particular, was also an especially friendly and informative organization.

From all these sources it would have been tempting to create an unwieldy slab of words. I must say I tried, but at Simon & Schuster, Bob Bender, ably assisted by Johanna Li, suggested, cajoled, or just skillfully snipped away with the editing pens to reduce that effect. Once they were done, Maria Massey was brought in to further sort out the pronoun forest, as well as unbraid the most tangled sentences. At an earlier stage, Becky Abrams was willing to pause in her writerly afternoons at the Old Parsonage to help with the initial idea for the book. Susannah Kennedy knows a great deal about anthropology, as well as most other subjects, though with her modesty she would probably demur. She helped me a lot: talking over ideas, pointing out the areas where not everyone would be quite as interested as the author; even once breaking an international flight to stop off at Heathrow for a much besought editing session. Throughout, the photographers, researchers, and scientists of Michael Marten's Science Photo Library did an excellent job of creating or collecting the pictures.

It would seem silly to write a book about families while holed up away from one's own. This is why Karen and I were willing to go to an IKEA superstore, even amidst the family-churning hordes of a Saturday morning, and get our nifty expandable dining table. You set it at one size (medium) for meals, and after, when everything's cleared, you tug it wider, and lay out your computer and notes and nonspill coffee cup—this is especially recommended with an investigative toddler about—and then, while everyone's busy with their day's work, you bang away at yours. The blessings of sharing this with Karen, my lithe and graceful wife, cannot be overestimated. At times when the stacks of notes got too large for the IKEA table, I retreated to a separate study, but even this was rarely in total isolation, as along with photos and Sam's collected bark sculptures, Sophie, now eight months, had developed a partiality for sitting under desks, gurgling at this and that, examining the world. Sometimes she'd look with hope at some distant—but reachable?!—intensely dangerous object, and in the process suggest text ideas for her thought-depleted dad. My own father was born before airplanes, the Somme, and long before television. Yet my floor-exploring daughter can easily have children who live in to the twenty-second century. Perhaps a snapshot of this odd epoch we're caught in—the last few years of the twentieth century—will be illuminating for her descendants to see.

This book was begun before Sam, our first child, was conceived. As it is finished, Sam is now three, old enough to walk me to that study, careful not to spill the coffee cup he's honored with carrying, and setting me to work with the cheerful injunction, "Be imaginative, Daddy!"

It's all that a man could wish.

prologue: 4am

THE LARGE DNA-CREATURE WITH TEN EYES AND FIFTY FINGERS DIVIDES INTO ITS COMPONENT PARTS AT NIGHT, EACH ISOLATED IN ITS DORMANCY AREA FOR REGENERATION . . .

the husband   The houseplant on the bureau is breathing down quietly over him as he undergoes the automatic eyeball cleansing, trachea widening, pulses of sugar-releasing hormones, and the bladder pressure changes we all experience at 4 A.M. His dinnertime food is slowly being digested, dozens of his brain cells are popping out of existence, and his abdominal muscle cells—valiantly used in the gym earlier—are slowly being rebuilt.

the wife   She is asleep beside him, with her brain busy in the mix of automatic data resorting and wish fulfillment known as dreaming, oblivious to the cosmic rays leaking in from the ceiling, which her DNA repair mechanisms are now actively defending her against inside her skin, ovaries, and other cells.

the daughter   Aged fourteen, and fervidly dreaming too, under an abundant acne lotion film, in response to recent distressing hormonal changes.

the son   Aged ten, and finally sleeping off the sugar blasts from the candy bars, popcorn, caramels, jelly beans, milk chocolate, and peanut-butter buttons consumed the previous day. The computer terminal on his homework table is turned off, but a central computer elsewhere remains actively awake for him, steadily listening to the Internet to collect his e-mail messages from distant friends.

the baby   An unexpected arrival, just ten months old. Half awake and gazing at the blue-and-red mobile slowly turning above; trying hard to concentrate, with brain cells hooking up in new configurations, there in the half-light from the hallway night-light; smiling at the sounds of the family dog wheezing away in the hall outside.

. . . UNTIL SLOWLY, AS THE HOURS GO ON, THE FAMILY COMPONENTS AWAKEN, TO UNITE GRADUALLY IN THE EATING CHAMBER DOWNSTAIRS.

morning

1

at the table

Yummy Yummy Yummy Yummy! says the father, reaching through the clutter of baby-food jars and serving spoons on the sunny kitchen counter, trying to find the right type of baby food this Saturday morning.

"Yummy," the mother chimes in distractedly from behind her newspaper, hoping for a few more moments with the Op-Ed page before she has to do anything too maternal and helpful right now.

"Mmm-my," coos the ten-month-old in his high chair, red plastic bib on and yellow drinking cup clutched in hand, his eyes brightly tracking the opened jar.

"In for daddy," calls the father, bringing forward the heaped spoon.

"Yep," murmurs the mother, turning the page.

WaaaaggggHHHHH! screams the baby, eyes wide, pushing the stronger new food aside and grabbing for his drinking cup's straw.

. . .

The baby puts his lips to the straw, and his tiny lungs pull upward, removing hordes of fast-flying molecules from inside the straw. Fragments of cold empty vacuum—tendrils of distant outer space—appear inside there. As a result, the entire weight of the earth's atmosphere, stretching miles above this house, slams down on the liquid, rebounding it vertically upward in the dark tunnel of the straw.

There's a click as the radio is switched on, and the baby reflexively turns to hear it. For a moment, by chance, the dial is holding on a foreign language station. The foreign sounds appear just a blur to the parents, but not to this baby. Each phrase is crystal clear. At this age even the most ordinary of human babies has a huge amount of extra circuits in its brain, and so the ability to register any of the several hundred distinct sounds in all of the world's languages—the clicks in Xhosa, trilled *r*'s in Spanish, growling *ch*'s in Gaelic. It's universal. Japanese babies can hear the *r* and *l* distinction their parents can't, while American babies can detect slight differences in Japanese, or in Hindi, that their parents are oblivious to. The skill only lasts a few months, so the baby traveler, not knowing which language island it's going to land on finally, concentrates now, struggling hard to grasp the voice sounds, and thereby keep the circuits that receive them solidly in place. Only when the dial's shifted away to a music station does everything change. Multitudes of brain circuits that would have stayed intact if reinforced start crumbling out of existence, they weaken, detach, are harvested by prowling housekeeping cells within the baby's brain, and then are dumped in the bloodstream for removal.

The baby looks at its dad for reassurance, and in a reflex almost impossible for a parent to avoid, the tiny muscles controlling the pupils in the dad's eyes suddenly tug wider. Males who don't have children rarely show this universal sign of interest. Women are different. Whether they've had children or not, they're almost certain to reflexively widen their pupils—it goes up an average of 3 percent in area—when a baby looks straight at them.

And then the dad goes ahead and wrecks it all by finally giving the baby that heaped spoon from the baby-food jar.

Some baby foods are fresh and wholesome, but many are simply the means for manufacturers to get rid of things they couldn't sell any other way. In the past it would have been harder to do this because people could tell if something terrible was added in. It would just sort of float around and you could look in and run screaming. But there's an excellent way to mask it all now. Insert a long polymer at the factory, one that swells when water is added, so much and so burstingly, that it stretches over the various added substances, making them impossible to distinguish from the food. As an extra bonus, the wondrously swelling polymer allows you to add in so much zero-cost water that it is often the main ingredient in baby foods.

One slight problem is that the polymer that does this swelling and masking tastes, to be honest, like wallpaper paste. But this shouldn't be too surprising, as it actually is the main component of wallpaper paste. To cover up the taste, tomato purée is often used as it is easily obtained, at conveniently low cost, from tomatoes that are too decrepit or just too bruised to be sold separately. The bright coloring also helps with the second problem, which is that the paste itself comes out a revolting gray when it's first mixed with the water.

To bulk it up, boiled and skimmed pigs' feet extract is often used, though in a pinch the scooped inner pith of discarded fruit can be added, too. Chalk is often added next. It tastes about as you'd expect (though it beats the wallpaper paste), but it is white enough to mask any of the gray gloop that shows through the tomato coloring. Vacuumed straight from a schoolroom eraser it would be too dusty to swallow. But with the polymers of the wallpaper paste, it mixes so smoothly with the water that is the main ingredient that it can be swallowed without a problem. Baby rice especially is color-masked this way, and it can be up to one-third straight, scooped-up white chalk.

The paste and water slurry now looks better, but it's still not a selling point to say on the label that chalk dust, pigs' feet, water, and paste are the main ingredients. Something more obviously enticing needs to be added to sell the product. Often that's meat taken from the animals we recognize as usual sources. But it's rarely taken from the *parts* of the animal we're used to. Cattle for example are largely fermentation chambers on legs and so have hundreds of pounds of mucus-lined digestive tubing inside. They need this to hold their gallons of bacteria and plant fibers in place until they are excreted. Such mucosal tubing would also be hard to sell if it were labeled, but baby food has often been exempt from any requirement to label the exact part of the animal its meat has come from.

Bowels accordingly are one of the more common meat sources used in

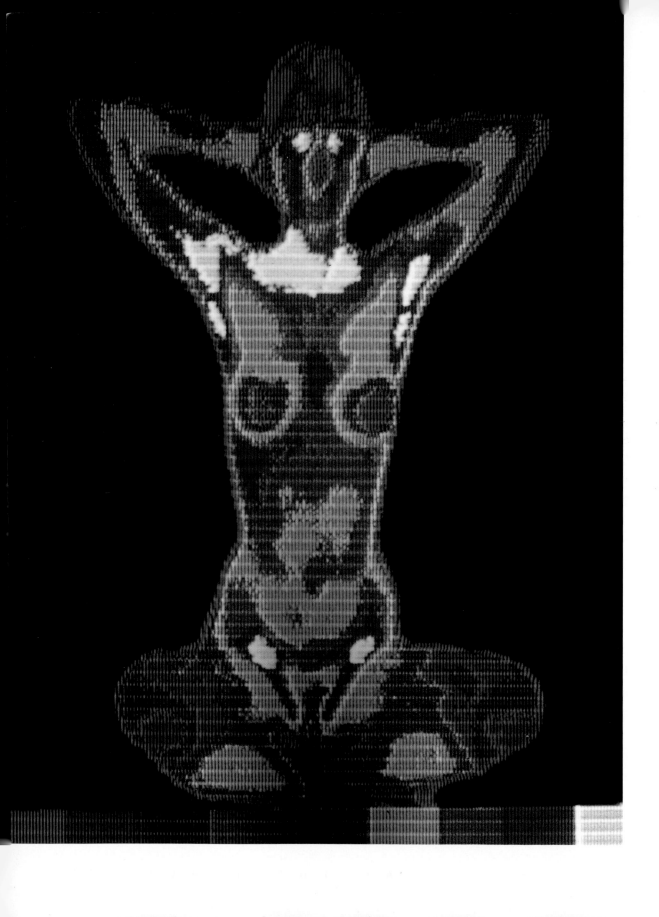

baby food. They are frequently put in great compression units with other hard-to-sell fragments—brains, testicles, and nostrils are especially common—and then they're all blended, squeezed, and cut into tiny cubes for mixing. If enough fat has been stripped off, the result can be labeled as "lean" meat. Sugar needs to be added to cover its taste. Kids wouldn't mind straight granulated sugar, but parents are fussy. Manufacturers, accordingly, often process fruit juices to yield a product that's chemically identical to ordinary sugar but usually can be heaped in without the dreaded *s*-word ever appearing on the ingredient list.

The mix is almost done, but it still lacks the right texture. The bowels, nostrils, etcetera, that came out of the compression units don't ooze with the connective fats that more normal meats would. Frothy chunks of animal fat get slopped in, to help along with vegetable fats as needed. A few herbs, an attractive label, some iron shavings to add mineral content, and there you are.

Usually that's it. But there are also some jars labeled as having extra ingredients which make them "suitable for the hungrier child." Sometimes it's processed cotton shavings or other cellulose pulp that goes in, other times it's just dollops of the dextrin glue used on the back of stamps. Both sound odd, but they're substances which swell exceptionally quickly once you mix them with water. Put the full water, chalk, bowels, pigs' feet, paste, sugar, fat, *and* stamp glue and cotton pulp mixture into a baby's mouth, and you can trust that he'll be left quite full.

The baby cries in horror at the next proffered spoon. The father is perplexed and tastes some of the spoonful, which seems pretty good to him. He only wants their tiny child, carrying all the parents' ancient DNA, to grow happily. But how can that happen if he doesn't eat?

Opposite: Thermogram of a seated woman. The hottest areas, seen in white, are the neck, ovaries, and armpits, where blood supplies are dense near the surface. Knees and nose are highlighted in the coolest blue; green is intermediate and often a sign of insulating fat layers.

The mother looks up from her paper, waking a little more—the increasingly sun-alert houseplants are spraying pints of oxygen into the room air, which helps. The mother and father consult. In a good marriage this is an easy task, for as they look at each other the parents are likely to see someone very much like themselves looking back.

With over 2 billion adults on the planet, the odds of any two pairing off are over 900 quadrillion to one. It seems an impossible task to ever get right, but there are a lot of characteristics we look for to make things easier. Some of the selection is purely physical. Eye color matches in married couples more than it would between random strangers, arm length matches and the length of ear lobes matches, and, roughly, attractiveness matches. It's only a rough match, partly because pretty women regularly end up marrying wealthy men, regardless of how the men look. (If he becomes less wealthy though, she quickly dumps him, as the statistics rather unromantically show.) But also the number of brothers and sisters you each had matches more closely than by chance, political opinions match, the amount of education your father had matches—even the likelihood of being a psychiatric case in future years matches.

About the only thing that's unlikely to match is the iron-loaded pigment called trichosiderin slipping out of the scalp in oil-lubricated fine tubes, producing what we know as red hair. People with red hair seem to hate marrying each other, and far more often than chance would suggest they pick someone with dark hair or brown hair or even no hair—anything to avoid someone with a red top. But it doesn't entirely work: red genes can be long-hidden recessives, and plenty of Americans with red hair have parents who aren't red-haired.

A final difference is more subtle. Spouses often smell differently from one another, but not simply because of the male reluctance to shower or bathe as frequently as women think that they should. Even clean humans generate a tremendous number of lightweight chemicals and these steadily float loose—a personalized invisible halo—into the surrounding air. Most are impossible to detect, existing just as a few isolated molecules, but a few hover more densely. These are the ones that enter the lower level of conscious awareness.

The effect is strongest on the immune system. It would be a great genetic advantage to select a mate whose immune system was different from your own, as that would give your kids an inherited boost through having more immune variations to build from. This in fact seems to be what happens. If women are given slightly used male shirts to sample, they almost always prefer those belonging to men with a different immune cloud from their own. When researchers go back and check, they find that the women's husbands or boyfriends also usually have a different vapor cloud. Which dating partners from the past were cruelly dumped simply because of the wrong immune system cloud?

A twist occurs when a woman has been taking the contraceptive pill. Probably because she can't sniff out other immune clouds as well then the tests show that she ends up feeling that the sexiest vapors are from men with *similar* immune clouds. Let her change contraceptive methods later though and trouble can arise, at least to the extent that we're chemical machines, for her vapor desires will take a swift 180-degree turn, and the same old spouse will be there. (Of course scent is only one of the factors that make up attraction.)

Over time, most of the main areas of spouse selection converge. The likelihood of drinking wine at the same rate comes to match in a married couple, the choice of breakfast cereal matches, and even the rate of going to art museums. People begin to resemble their spouses, and a couple are likely to go gray together, which is perhaps unsurprising as they were likely near the same age to begin with. The average is the man being three years older, a figure that holds in the United States, most of western Europe, and even in many hunting tribes.

The most curious change is in IQ. Here there's no bland convergence towards the middle. IQs match only roughly at first, but then—maybe it's those art museums you've been dragged to?—five or more years into the marriage, the score of the partner who had the lower IQ starts to rise.

Soon the older kids are called down. Only the ten-year-old boy gives signs of life at first; his fourteen-year-old sister is inert, still snuggled in her sheets, trying to delay the arrival of full consciousness. Early in a night dreams are brief, tetchy little things, twenty minutes long at most, and spaced at tedious, interminable, hour and a half intervals. In the morning though it's different, and if a teen can keep herself in bed and dozing, can squeeze her eyes closed and ignore the running dog and crying baby and her parents trying to locate the magazine section of the newspaper, attached electrodes would show her dreams popping into awareness almost continuously. Several neurotransmitters are sent out at lower than usual levels in the brain during dream sleep. Logical reasoning requires them to be present at high levels. That could explain the surge of pleasant—yet quite illogical—wish-fulfillment jumps we experience during dreams. Her body has been preparing her for waking—our blood volume and body temperature and adrenaline levels automatically start rising from about 6 A.M., in readiness for that—but with a little effort that tendency can be overridden.

The rush of dreams at the end of the sleep cycle, incidentally, is one reason jet travel from west to east—say, from California to New York—is so much more exhausting than the reverse. Since the eastbound travelers have to wake up at what seems an earlier time, they won't get as much time for those rich, world-orienting dreams at the end.

The boy emerges from his bedroom, and bangs with typical ten-year-old's delicacy on his sister's bedroom door, telling her that she doesn't have to bother about her beauty sleep, because it's not going to help anyway. She tries to ignore him, but her dreams are fading anyway.

They might deny it, but a brother and sister are more similar than anyone else in this home. The reason of course is that they have two parents in common, and all siblings share 50 percent of the same genes, on average. This is a feature they share with no one else. Everything from IQ to blood types to the immune vapor clouds and of course parental income levels is a close match.

More calls rise up the stairs, and although at first the girl shouts down with world-weary disgust that she's coming, the kids finally do descend, carpet electrons skidding underfoot to mark out their first house-crossing commuting trail of the day. The two kids bring computer games and colorful tracksuits and the joy of fresh life, yet also—quite regularly, in even the cleanest of well-off suburban homes—they bring something else, something quite small actually, cute to those who study it: an entire parallel family to the humans now assembling around the breakfast table. Actually it is more than a single family, it's family after family of little exploring creatures, strolling peaceably across the human foreheads, fond especially of the teenage girl's skin terrain, where they poke about for any leftover acne cream, or even just leftover mascara, which might make a nourishing meal. They can easily transfer to the boy and parents for further explorations.

These are the demodex mites. They're nothing like the awful, visible lice one can get from unclean conditions, for demodex thrive in clean homes. They're impossible to see with the naked eye, being only just the smallest dots even under a magnifying glass. But under an electron microscope they look like clomping mechanical cars, the adults each having eight pudgy little legs, on which they waddle slowly between their hair follicle homes. There's another, quite different, mite world we'll see later in this house, but the demodex are the most mobile and the only ones that make it to the kitchen. Virtually wherever researchers bother to examine, on our foreheads or near our eyes—a single eyelash will usually do—there the demodex are.

At first they can't really survive well out on the exposed outer reaches of the face. That's why the baby demodex, the cuddliest newborn pudgy ones, stay deep within the hair follicles on the face, nestling tight beside the soothingly warm root of a hair. Human hair can't grow without some liquid around its roots, and the baby demodex simply cascade upward on that, in a great sloshing water-slide ride. Most of us have at least a few bacteria for them to scavenge there. If the growing demodex has been lucky enough to land in the follicles of a teenager with acne, then there are supplies of nice, succulent, *propriabacterium acnes* and related bacteria waiting.

For about three days the demodex baby gorges and grows pudgier. Finally it changes form—an eager college graduate, putting on its best suit for the first day on the job—and then, for the first time, takes the final step forward, and it's out of the cave hatch and alone, tottering slightly, on the human face itself.

It's not quite isolated, for off in the hazy distance, there are other demodex graduates, also just emerging from their protective hair follicle bases. Amid the tentative newcomers are a few grizzled oldsters, cynical veterans of the tough life outside. The average demodex strolling across our faces is about ten days old. What it seeks—the stuff that's even better than the acne gunk inside the follicles—is all around. These are the fields of bacteria found on all human faces, generally quite harmless, especially the tremendously nutritious *pitorosporum ovale* sort. The walking demodex lower their pudgy heads and get to work eating.

A forehead-touching morning kiss is one way they spread within a family, but indirect methods are more common. The bathroom face and hand towel is ideal for moving from one family member to another.

In time though, as happens to all travelers, the lure of home becomes too strong; the isolated demodex grazers stop their feeding, lift their heads, and then, after a careful chemical sniffing of the environment, turn to go back. The ones still migrating, perplexed, on a bathroom towel are sunk, but the demodex safely on our faces have a good chance. It's usually about sixty hours since they left home, so the demodex that are heading back on Saturday morning will have been on the go since Wednesday afternoon. If there are too many eager youngsters at the original cave exit, or if it's blocked for some other reason, the pudgily returning demodex will have to find another home, but urgently now, for only at the edge of a follicle cave opening can they undergo their final growth, to become full adults.

The females stay at the hair opening once that happens, but the males don't have any time to waste and head straight back out. Now, for the second time in their lives, they have to wander over the landscape of the face. But this time they're looking for another follicle opening, one with a female, and they have to find it fast, for they're almost twelve days old now and don't have much longer to live. They trudge from one cave exit to another, repeatedly turned away by the other families already living there. Many don't make it and end up collapsing, alone, unmated, on the surface of the face, to be digested by the very bacteria fields they fed on, or possibly washed off onto a shower floor, if there's an especially thorough scrubbing by a hygiene-frantic human later this day.

The females have a longer life, and can wait up to three or four days at the hair follicle edge, for one of those exhausted waddling Romeos to appear on the far horizon and hurry closer through the bacterial fields, till there's contact, a discreet repositioning of bodies, and then the fertilization that will allow this particular genetic line to go on.

Wonders in a kitchen sink. At $^1/_{500}$ the thickness of a human hair, ordinary detergent bubbles are one of the thinnest substances visible to the naked eye.

Every human family gradually builds up its own distinctive population of walking demodex families. When the husband and wife first met they each had different subpopulations, from their own families. Only during the first few months, when they were dating or dining or tentatively living together, did those separate populations merge. Sometimes the result is true cross-breeding and a new unique group is created; sometimes, if one of the humans supports an especially hardy population, it's similar to someone who takes over a new relationship by insisting on carting in all their old books and old friends. The demodex newcomers out-eat, out-walk, and generally force away the previous ones on the spouse.

When their daughter goes off to college she'll carry a rich sample of the parents' now long-established population with her, and later, if not at her first Thanksgiving back then certainly by the next summer, she'll have fresh new-comers to offer, picked up from her foray into distant dorm rooms or rented apartments. But that collecting is only fair, as she'll also have left members of her parents' unique breed out there, hitching forever on the lush forehead and follicle worlds of roommates and boyfriends.

By now, waiting for breakfast, everyone's desperate for food: our brains run at a power rating of 20 watts, and this family's brains haven't been replenished all night. Murmurs, exasperated groans, and then the worst of all—whining demands from the kids—break the silence. The baby is lifted from its high chair, yelping at first at the floor passing so immensely far below as the great voyage starts. Its head tilts slightly to its right, which is the ideal reflex twist should the father continue the movement, to slide it into the much-sought docking position on the parent's left side. Here the adult's reas-suring heartbeat is close below the surface. Babies are usually held there, and it doesn't matter whether the parent is left- or right-handed. Renaissance paintings and ancient vases alike almost always show babies being held on the left. The only klutzes are first-time dads who often miss the baby's sug-gestive head twist, but even they learn in time.

Still holding the baby, the dad opens the refrigerator with his free hand, releasing a blast of cooled air, which the baby sniffs. A family of two adults and three children will weigh approximately a quarter ton. In one year they will consume about 3,000 pounds of food and 900 gallons of drinks—this is why parents get that sickly grin when asked if they enjoy grocery shopping—so the refrigerator needs to be powerfully energized to store it all. Electrons are pulsing along the cable at its back, sliding forward a mere inch or so at a time, in response to the voltage charge hurtling through from the power station. If the station runs on coal, oil, or gas then this family is getting its food cooled by remnants of carboniferous forests and microscopic life-forms that waited 300 million years to yield up their final energy in each fragment of cooling gusts here. If it's a power station running on nuclear power, then an energy source older than our sun is being shunted through these cables into the house, for the uranium in these stations is simply the ash of earlier exploded stars.

Before such efficient cabling, chunks of ice were used to keep foods cool, but this wasn't easy to get. Mughal emperors of the sixteenth century tried, but when your headquarters are in Delhi this is a losing proposition. Riders started out from the Hindu Kush mountains, possibly with a certain amount of confidence if they hadn't done it before. But however well they packed away the ice chunks in their saddlebags, crossing the plains leading to Delhi the heat would bake and soak and generally blast into the saddlebags. In some miniature paintings from the time you can see the emperors and a favored few of their court with the result: tiny little ice sorbets. A somewhat more promising method was the New England system of carving up frozen ponds with horse-drawn iron saws, then coating the ice blocks in sawdust, and storing them in vast houses with double walls and insulated roofs. In the early 1800s, New York families who could afford it would sign contracts for delivery of 1,200 pounds of ice—eight pounds lugged freshly up their stairs each day—to get them through the summer. Ships could be similarly insulated, and by the mid-1800s New England ice was being shipped to the Bahamas and even—the Hindu Kush horse system long abandoned—to India.

Refrigerators profited from the ingenuity of James Harrison, an Australian printer. In 1851 he was cleaning blocks of metal type with ether and noticed how cold the metal got in his hand as the ether evaporated. He developed a machine where entire tubes of metal were cooled this way. The problem though, aside from the great size of the machine, was that the chemicals used in the coolant tubes tended to smell awful, and so long as that was the case home refrigerators had to be big devices, only for the rich, with the reservoirs for the tubings often kept in a next door room.

There was no solution to coolant problems, until finally, in the 1930s, one determined Du Pont chemist devised a mixture of chlorine and fluorine on a tough carbon core. The name it was granted accordingly was that of a chloroflurocarbon—he had just made the world's first CFC.

As the sounds of the family grow louder, the baby, having been put down on the counter, is enjoying its exploring too much to listen. Further along the counter a bowl of fresh apples are also talking among themselves: spraying out ethylene gas in simple data streams to coordinate their ripening schedule. (Fruit is often picked before it's ripe, and intentionally sprayed with this *"ripen quickly"* signaling gas when it's time to be moved to the stores.) The baby leans forward, and its moist breath cloud scatters aside the tiny ethylene puffs, throwing the fruit's ripening suddenly off schedule.

The marauding baby turns back to his father, who's still looking inside the refrigerator, helped by light from the bulb inside. Inside that bulb is a filament glowing at several times the temperature of a blast furnace. This sends tungsten atoms furiously bubbling up, but the whole process is insulated so well within the oxygen-empty bulb that hardly any heat escapes. The ten year old is likely to have repeatedly flicked the door open and closed during the week to examine the light's mechanism, but although parents complain about imminent bulb burnout, there's little chance of that. The notion that lightbulbs use up extra energy each time they're turned on and off comes from confusing them with fluorescent tubes. Those do undergo a sparking jolt each time you fire them up, and it is more efficient to avoid turning them off if they're going to be needed again soon. Ordinary incandescent bulbs, though, in the hall or this fridge, are independent of flicking hands, just quietly evaporating until their filament narrows and decays enough to snap.

The baby can squeeze its eyeballs to get a sharper view of this illuminated box, but it still doesn't see what we do. The edges of the refrigerator and its parents faces and hairlines will be in sharp focus; it's the bits in between, not yet fully comprehended because of its still incomplete brain, which remain slightly fuzzy.

The dad is having trouble finding what he wants. Men, on average, are exceptionally obtuse at processing information in three dimensions. Boys do worse than girls on tests about it at school, and often have problems even with simple inversion—That's probably why men are turning this page around while women are managing it this way just fine. The thick connecting cables of the corpus callosum that shuttle information between the left and right sides of the brain are usually less well developed in men; scans using nuclear magnetic resonance imaging suggest that a man faced with a three-dimensional packing problem is more likely than a woman to work primarily with just one side of his brain when he starts.

He's also likely to have a different search style. Women generally remember their way around paths on a map by memorizing particular landmarks. Men tend to skip those useful reminders and more often rely on pure guesses of relative distances and angles. In the refrigerator that might mean the wife knows to look for the fruit juice past the milk and behind the jelly; the husband is more likely to vaguely remember it as being up and on the left.

There also are strange seasonal variations, for higher testosterone levels sloshing inside exacerbate the problem. Men almost always do better at three-dimensional orienting in the springtime, when their testosterone levels are at an annual low. Men might be tempted to take some comfort in the fact that at least the total volume of their brains are greater than that of women—the difference is about 10 percent. But alas, such reasoning would only be a sign of these debilitating male limitations. Men generally weigh more than women, by that 10 percent or more, so of course their brains are bigger. A whale's brain is bigger still, but this does not make the aquatic creature a wiser beast. When you factor in women's smaller body sizes, their brain sizes are proportional to those of men, or indeed a little larger. Suffering all these handicaps, it's only with extra effort that the father finds the drink he was looking for and brings it back to the table to his waiting wife and kids.

HOW typical is this family? Despite America's recent high divorce rate, the most traditional arrangement remains common: about 75 percent of children still spend most of their time in a home with two parents. For a brief period after World War II the rates were even higher, but that was a historical exception. People didn't stay married for very long in previous times, not so much because of divorce but because adults died at young ages. Life expectancy in early colonial America was about thirty-five years, and in much of Europe it was less. Marriages were so short—in 1880s America they averaged just twelve years—that families were repeatedly made of half brothers and half sisters. The great number of stepparents in nineteenth-century fairy

A kitchen scouring pad. The yellow stripes are the nylon ribbons that keep it scraping effectively.

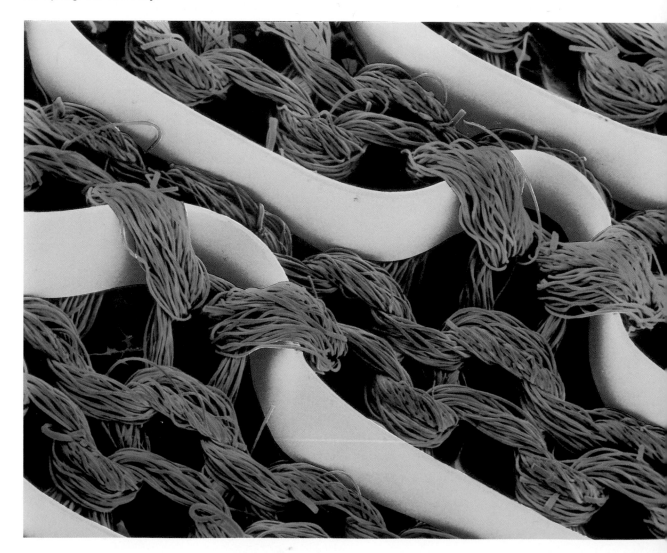

tales is a relic of this time, when almost everyone could count on a new family configuration at some point in their lives.

Right now though, every family member except the baby is sharing the first breakfast drink—a liquid which contains embalming fluid, varnish solvent, vinegar, and nail polish remover . . . and a certain amount of real orange juice, too.

The problem is entirely due to that last ingredient. Orange trees are so good at soaking up groundwater and transferring it to their little dangling fruits, that it costs a lot to ship an entire orange to market. It makes much more sense to squeeze the oranges near the place where they're harvested, and transport only the lighter-weight concentrate. When a local canner receives the shipment, it's accompanied by a manifest stating that, say, 12,000 gallons of water have been extracted. This is where the problems begin. An honest person would say yes, I must put back the 12,000 gallons and not a single drop more. The reason I can't add any more is that if I did, what I've received would go further, and be divided into more cans and jars, and that would give us easy excess profits, and that would be unjust.

Since not all canners are this saintly, extra water is often added—a few thousand gallons, often thinning it down by about 15 percent. But chemistry is an unforgiving science, especially when keen-eyed government inspectors are on your back. The water that's added back in can't be just any water. Natural water has a different isotope mix from what accumulates inside a fruit tree—inspectors can easily detect that—so the canner or bottler has to order what's politely called "pulpwash." This is what you get by taking the piles of tattered oranges that have already been thoroughly crushed and heaving them into waiting baths. Orange fragments scraped up from the surrounding cement floors can also be added. The mix is left to soak, the rind and pith will partially decay, and so a little more substance can be squeezed out later. That's the pulpwash. It's not really juice, but when shipped to the final canning plant it is added to the orange-colored cold soup. Certain popular brands contain up to one-third of this recycled pulpwash, even when they're labeled as "pure" or "freshly squeezed" orange juice.

Pulpwash on its own tastes terrible, so sugar has to be added, then some acids and a little acetone (the active chemical in nail polish remover) to give it some tang. When you do that though the acids start reacting with each other, so there has to be some vinegar to slow down that process, but you don't want to slow it down too much, so some ethyl acetate (varnish solvent) goes in. To keep it all from breaking apart, some of the embalming fluid formaldehyde—or a chemical near cousin—gets added as a final salvaging touch: the chemical is ideal at forming tight linking groups between proteins, be it dissolved bits of cadavers, or the more palatable pulpwash. Rail freight moguls who owned excess refrigerator cars from the Chicago stockyards began the fashion of delivering fresh fruit from warmer climes by advertising this concoction as orange concentrate.

The first sips of the refreshing orange, pulpwash, varnish solvent, polish remover, and vinegar mix go down, swallowed cautiously by the teenage girl, who's suspicious of any food substance that might contain calories, but rather less decorously by the ten-year-old boy, who's trying to glug in the way awesome beings—or at least the adults in those beer commercials on TV—should. Around the table each person's detoxification system immediately starts working: acetyl groups of atoms are prepared in everyone's liver and are added to each of the alien molecules as they come through. Without this process most of the food additives would get stuck, suctioned into our bodies' fat. With this molecular addition though most become soluble in water, so that they're easily led through the kidneys and on to the vast inland sea of the bladder for safe disposal later. It works fast: some of the quite toxic aluminum which the oranges pick up from the soil around their trees—aluminum is the third most abundant element in the earth's crust, so there's some spattered almost everywhere—gets fixed with this acetyl group so quickly that it's bobbing in the bladder of everyone around this table within twenty minutes of the first swallow. (Two organs get special protection against toxins loose in the bloodstream even before the acetyl groups get to work: one is the brain, and the other is the testicles. Unlike other organs, they are blockaded from all ordinary foods and blood travelers. They survive almost exclusively on a diet of pure sugar and pure oxygen, because very few other substances can pass their tight capillary barriers.)

A single chromosome location controls the main detoxification route, but it does not operate in every family at the same efficiency. In one unfortunate British company, twenty-three workers died of exposure to benzidine from the dye works they were tending. Almost all came from families who were especially slow acetylizers. The other workers had absorbed the same amount of benzidine but, possessing a luckier inheritance, had simply managed to float it away. In Asia a relatively high number of families are fast acetylizers and so are relatively safe from the estimated pound or more of completely indigestible colorants, stabilizers, preservatives, packaging gases, dyes, emulsifiers, anticaking agents, and miscellaneous metals a supermarket-rich diet will produce in a year. Most Europeans and Americans are only medium or even slow at the all-crucial acetylation, and so are exposed for minutes longer. One quirk which works in the opposite direction is that most westerners are quick to transform wine or beer when it first flows within their bodies. The reason could well be a genetic peculiarity that spread through the population. In Japan and China by contrast, more people are likely to lack the gene that ensures a fast breakdown. Sudden flushed cheeks—and a disorienting quick rush to the head—can result.

Now the mother doles out further health insurance: a vitamin pill to each family member, the guarantee that even if she hasn't been able to do all the shopping for fresh, nutrition-sopping vegetables she knows she should, she'll still be providing for her kids. But vitamins are actually tiny molecules, so the potions don't really have to be big as what's handed out now. Yet who would believe in the power of tiny pills? They're bulked up accordingly, and almost all of their volume is made of entirely useless fillers: there's sand and chalk and talc and, to bind the whole thing really tightly, steamed extracts from pigs' feet and other animal bones. Each ounce of actual vitamins the family swallows over time brings with it far more of the fillers. When the pill finally lands in the stomach, the vitamin at its center has so many of these extras to push through that the vitamin only slowly dissolves loose. About half of the most popular vitamin pills stay in such large granules that they can't be digested at all, and simply move along the digestive tract unchanged. The other brands are only marginally better, for unless a family is suffering from a severe food shortage most of the vitamins are immediately sieved out, leading, as pharmacologists put it, to Americans having some of the most expensive urine on the planet.

A century ago no family felt the need for vitamins, as food was thought of as one big undifferentiated thing, and all that counted was how much you got. This utopia ended with Casimir Funk, the Polish chemist who, with others, first recognized the importance of certain trace substances. He modestly resisted the suggestion of calling them Funkians, and instead selected the label of vital-amines, or vitamins for short. Had Funk been less modest, kids today would be encouraged to drink up their Funkian C, or take extra Funkian capsules to be sure.

The surface of these pills is often more potent than the molecules inside. Angina patients given placebos colored yellow rarely show much improvement, but when the placebos are red they produce a strong effect: enough to change blood pressure for days. Vitamin pills are often soaked in red dye to match such psychological (or placebo) effects.

The family discusses the next course, and also starts to plan the day's mall visit. A video would show a strange spectacle at this point, for even in the busiest of conversations we remain utterly still and silent about 35 percent of the time. The patterns of conversation are also odd, with most people repeating what they say in various forms, insistently, more than once, vocal chords whapping together, over and over again, until, finally, someone repeats it back to them and they stop. Little head nods from our listeners make most speakers go faster; if the listener is sitting with arms crossed most speakers go slower. The whole process usually begins with the speaker-to-be glancing away or down for an instant, as he or she seems to need this moment to plan the initial phrase. It's surprisingly hard to start talking if you're looking directly at someone.

What we see of each other while we're talking is strangely distorted, too. Everyone you're looking at first gets turned upside down and miniaturized, to fit on the inside of your retina. Then, as these images are shuttled along the optic nerve deeper into the brain, the truly surreal transformations begin. The family picture doesn't stay as a tiny reversed movie screen, but instead is broken apart into separate processing areas for hue, edges, movement, and the like. A vast number of isolated data bits, scattered throughout the multibillions of quick-firing wet brain cells, is all that remains of each person we briefly scan.

The son asks how much spending money he'll get for the trip, and the mother hesitates, trying to remember what they had agreed. What makes it hard for her is that an incorrect memory feels remarkably similar to a correct one. Dredging up either a true or false memory increases blood use in the part of the brain where stored data is accessed. The only difference is that when we access a true memory we're also likely to switch on the parts of our brain where the physical sensations of the true event took place. But short of bringing a huge brain scanning machine to the breakfast table to try to recapture that faint memory, how is a tired parent to be sure?

Letting the kids cajole, quiz, or even just speak around the table has not always been allowed. Erasmus, writing around 1500, was considered a dangerous innovator when he suggested children might be allowed to say something at meals "when an emergency arises." In wealthy Prussian homes in the late 1800s it was permissible for a father to punch a child with the back of his fist for speaking out of turn. This is not quite the American way, and tolerance for children expressing their views has exasperated European visitors at least since the Revolution. Some attitudes are surprisingly persistent. Children have been observed with their parents in playgrounds in Italy and in Germany. The Italian parents would hardly ever interrupt or hit their children over an hour's play; the German parents were going at them almost every minute.

Everyone's caught up in the family-linking discussion, except the red-plastic-bibbed prime member, isolated in his countertop exile. He's been watching just as eagerly as everyone else, but all the air bursts from the other mouths are hard to follow. His solution is easy enough. He concentrates, then creates a special nitrogen-vapor blast from his own mouth—the *scream*—to draw attention to the injustice.

The dad half fills the plastic cup with more juice, then passes it to his youngest. The baby smiles as he accepts this, taking it as his due, knowing he can now relax; triumphant, once again, in his essential parent-controlling skill. The vaguely orange mixture pours down his tiny throat; deep inside his body, the first quick detoxification starts biochemically whirring away.

And then, like a fool, the father asks his daughter if she'd like some more, too.

A creature furious with adolescent agony now emerges. Does the dad realize what he's done? she asks. Does he—does anyone in this whole family?—realize that she's been trying to keep to a diet, a sensible one, and not load up on this (a furious look at the proffered pitcher) this, sugary *junk* they insist on. Do they *want* her to not fit in any of her clothes? Maybe they've forgotten—hah, as if they'd remember—that it just so happens that she has a very important date coming up? No, she says bitterly, whipping open a magazine to flick through, already staring away from them all; they probably don't. They probably don't care. They . . .

The baby is enthralled, but everyone else knows it's best to look away for a while, and not argue back. The wife closes her eyes, and rubs her forehead. This is soothing not just because you no longer have to see your offspring, but because it helps change your brain waves: slipping from the jerkily uneven short pulses of ordinary attentiveness, to the smoother, more massagingly deep rhythms of alpha wave activity. Adolescence is not a new ailment. Aristotle recognized its symptoms 2,350 years ago, in the second volume of his *Rhetoric:* "They are hot-tempered and quick-tempered," he wrote of the young Athenians around him; they "can't bear being slighted, and are indignant if they imagine themselves unfairly treated"; they also "think they know everything and are always quite sure about it."

What makes it worse now—an exquisite prolongation of the agony for all concerned—is that it lasts so much longer. In Bach's choral accounts book of 1744, boys were listed as singing soprano at age eighteen. Girls also often reached puberty that late—even during the American Civil War puberty did not often occur until age sixteen or seventeen because of terrible nutrition. Poor countries today are similar. Among the Kikuyu of Kenya, for example, girls still only achieve menarche at age sixteen. The rich countries are the ones that have changed, with puberty beginning years before kids—these aspiring mini-adults—can leave home. Tension, and a drive for independence ensue.

You can make some predictions about how bad the battles are going to be, at least roughly. Families that talked a lot when a child was aged twelve have the best chance of ending up in that group of blissful households—57 percent of the total—where teenagers report actually enjoying their parents during adolescence. Girls who are especially pretty when they're eight or

Top: an immature egg, in an ovary where it's being nurtured. *Bottom:* the fully matured egg tumbling loose. Mood- and body-affecting estrogens are released from the ovaries as well.

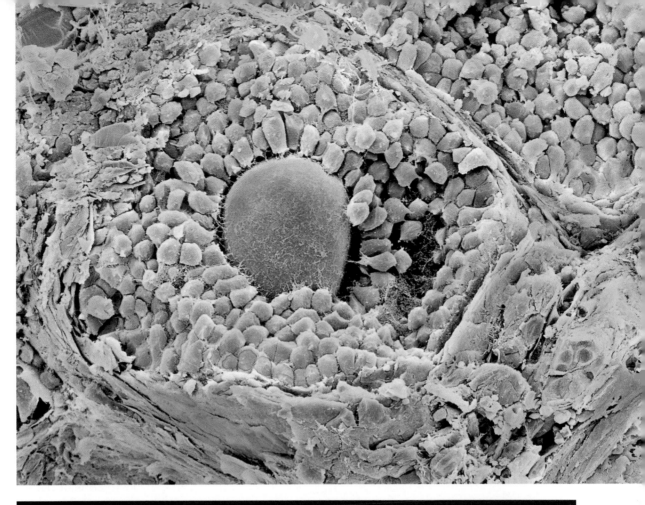

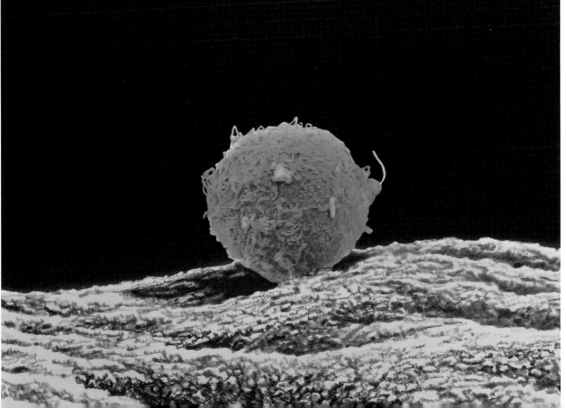

The chunky slabs of an ordinary eggshell. In a fertilized egg the slabs would steadily thin, giving calcium to the fetus for its skeleton and weakening the shell to make exit easier.

nine years old rarely end up in that lucky 57 percent, probably because they have more to lose from any change than their plainer schoolmates. And if a family's income has suddenly gone way up (or way down) when the daughter is starting adolescence, then she's especially likely to feel aggrieved, as she won't have a clear standard of family background to latch onto with her friends or boyfriends.

As the daughter flicks through her magazine, chlorine fumes from the bleached wood fibers billow unseen around her head. The father tries to return to his primary food-gathering role. But the hormone that's normally released from the brain to deal with stress, ACTH (adrenocorticotropic hormone), is likely to have overshot its mark, and several dozen nanograms will still be cascading down in a reverberating response to his daughter's flare-up. This can upset his immune system for hours, making him more susceptible to any cold viruses or other microbial assaults later in the day. The broken sleep because of getting up for the baby during the night makes it worse. When brain waves are measured in sleep labs, people with the most delayed responses from their immune system cells are usually the ones who've had the most interrupted deep-sleep brain stages. Only if the dad manages not to get too aggrieved—it helps to recite Ogden Nash's promise, that some day one's own offspring will get the delights of seeing their own children adolesce—might the peptide flow shut down quickly enough to avoid that need for imminent cold remedies.

A forced shrug of the shoulder, and the dad surveys the counter again, till he locates the yeast-pumped grain-dense sticky balloon known as bread. Over its surface an extruded polyethylene wrap has been plopping down tiny droplets of oily plasticizers and lubricants. Fifteen percent of a plastic wrap's volume can be such extra chemicals, put in to keep it smooth. In the family car parked on the street outside, the buildup of the day's heat makes enough of these smoothing chemicals evaporate so that they are detectable by sufficiently attuned noses later. Here most of the molecules simply pool up on the pitted grainy surface and slip off as the bread itself is pulled out. Fatty foods like cheese soak up even more. Most of the transferred molecules are harmless and easily detoxified, though some seem similar enough in shape to female estrogen hormones that if the wife ingested any amount when she was pregnant, the baby might have problems in his sperm production cells once he reaches adolescence.

White bread used to be the product of choice for the wealthy, as only rich people could afford the chlorine bleaching process and the sieving out of all the burlap, dirt, coarse hulls, and other extras with which peasants' bread traditionally came equipped. In France, for example, purchase of the fluffy white baguettes was a sign that a family had made it in the city or at least was not going to be tied to a farm any longer. Fashions changed when it was found that, although the burlap fibers and occasional powdered dirt could be left out with no harm, the dark hulls and other extras actually were better than the white stuff. Nutritionists regularly rediscover this fact, but does anyone listen? Even in 400 B.C. Hippocrates, the Greek physician, was haranguing his patients to eat whole wheat bread "for its salutary effect upon the bowel," but the records show Greeks of that generation and later still preferred white bread whenever they could afford it. Today's nutritionists also have to reckon with the stubborn resistance of bakers themselves, who generally try to discourage whole wheat bread sales. The oil in its wheat germ easily goes rancid, which means whole wheat loaves can't be stored as long as others. If you buy your bread very fresh though there's an unexpected boost. Most of the alcohol produced by the yeast that makes bread rise will have evaporated in the baking, but straight from the oven a certain percentage still remains.

The dad pulls out the first slices of bread, and deposits them in a shiny metal box, attractively glowing for the baby's interest. Inside, transformed starch molecules near the bread's surface are leaking out as dark ooze. This is the dextrin that makes toast easier to digest than unheated bread, as well as providing the crusty brown color. Toasting your bread at home would have been fatal to try in Cambodia when the Khmer Rouge took over, for all private cooking was outlawed. Every family had to eat at a central mess hall. If you tried to cook separately, and anyone saw you or informed, you would be killed.

Here though the dad is safely busy—in a world where parents have to work, isn't it fair to have fresh pancakes at least one day a week?—and he swoops the baby down to the floor. Eggs are brought out from the refrigerator. When he sees that they're running low he prints a note to get more and adds it to the crucial family communication device which architects repeatedly forget to supply: he sticks it up with a little magnet on the refrigerator door. Invisible curves of magnetic force swoop into the room's air, generated by quick spinning atoms inside that magnet. The iron atoms were created in a slow buildup over aeons in distant stars. The magnetic lines streaming through the dad have little effect on his body, but the ones contacting the metal door hold the paper on tight. Ancient Greeks thought that natural

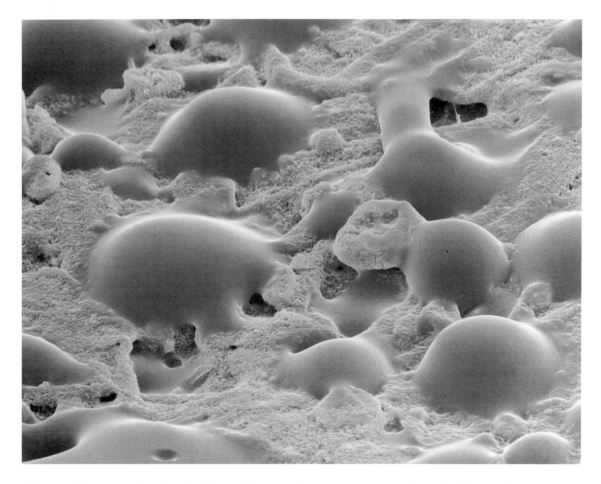

These igloos on the back of a yellow sticky note contain liquid glue: a few pop open each time the paper's pressed down, whence the repeated stick—at least till the last bubble's burst.

magnets came from an island entirely composed of the stuff, which anyone sailing near would be pulled toward and lost forever. Such an island has never been found—though who could report back if they got there?—and modern magnets are made by exposing appropriate materials to powerful magnetic fields supplied by a surrounding electric coil. The other portable memory holder on fridge doors—the ubiquitous little yellow sticker—almost didn't make it, for the 3M scientist who discovered their glue nearly discarded the formula when he realized what a poor adhesive it was. Only when he saw that this made the stickers uniquely useful for other tasks—marking his place in a church hymnbook was apparently their first use—did the promotions start.

The dad finishes his note and gets ready for the cooking. But which way should the egg be held? In *Gulliver's Travels,* two great kingdoms fought over the issue of whether it was the pointed or the widened end that was superior. Jonathan Swift, the author, thought he was being satirical, but this is simply because his comprehension of the buoyancy properties of the avian yolk was inadequate. When eggs are stored with their narrow tip down, then the air space inside will be in the wide open area under the spacious dome at the top. The yolk is barely buoyant in the dense medium of the egg white, and can't float upward. Only if some crude individual insists on storing the eggs with the pointy tip *up* will danger ensue. Then the air space is on the bottom, and the depth-bobbing yolk can easily bump against it. Oxidation or at least a partial drying out—not to mention a less attractive taste—will soon ensue. In any event, the egg's ovoid shape means it's unlikely to roll off the counter while it sits there waiting. Spherical eggs would easily roll away from a nest. Ovoid ones are safer, and in fact the eggs of birds that nest on cliff edges are often the furthest of all from a sphere, giving them the tightest rolling arc.

The shell splits open as the egg is cracked against the mixing bowl. The fresh yolk tumbles neatly over the mixing bowl edge. Buyers of free-range eggs might pause at this moment, musing on the little oval so recently in their palm; pleased that their free-range purchase, despite its whopping price, gave this health-bursting little thing and its parent a life, a real life, that was good and wholesome, and, well, *free,* there on its open farm.

It's a noble ideal, but not always quite true. For free-range eggs are merely covered by a law that says the chickens must have access to open territory. It doesn't say whether they have to be encouraged to actually go ahead and use it. Since it's no fun taking care of hundreds of chickens running around, manufacturers often try to make sure that the chickens stay inside. It's not really very hard; there's no need for little stubble-chinned rooster guards with Uzis. For chickens originally evolved in the tropics, and only if there's a space with trees and shade and running water or other Edenic accoutrements will they run outside. To keep chickens inside you just have to make sure none of those items are nearby. A tarred-over open surface with a chainlink fence around it is pretty good, and if a dog can be chained nearby or a narrow cement walkway built as the sole path leading from the main chicken roosts, you're doing even better. The "free-range" chickens can end up cooped inside just as much as ordinary factory-farmed birds.

The legal trickery is deft, but sometimes there's a way to tell when you've been cheated. Chickens that spend all their time inside often need artificial colorants added to their food to make up for their limited diet. The result is a giveaway unnaturally bright yellow color in the floating yolks. It's especially noticeable in the winter months, when even more food additives are needed. Embarrassingly lurid golden yolks can appear where an unpracticed farm manager overdoes the Technicolor dose. (Butter is often similarly colored, for although summer butter is naturally yellow, winter butter would naturally be pale. Here there are fewer giveaways though, because the yellow colorants in butter are easier to control.)

The chickens that grow inside the breeding houses are sorry-looking birds when they're finished. There are all the injections and strange feed products—recycled egg cartons and waste-drenched straw and even asphyxiated bodies of unselected males have been observed being ground up and used as feed—and because the chickens are bred to be scrawny, and are packed closely together, they're often pretty bruised once their lives are up. If only there were some way of getting rid of them, for cash, that wouldn't involve anyone from the general public glimpsing the shape they were in. It would have to be some very vulnerable creature that fed on these chickens, of course; one that had no choice about the matter, and might even be strapped down for the feeding ordeal.

The dad glances at the rows of yummy chicken baby food, stacked waiting to be used at the counter's far edge. But where is the ostensible consumer of those jars now?

The baby couldn't really be expected to stay still there beside his dad's legs on the floor where he was put down. Straight ahead there's the

The windblown shrubbery of an ordinary fungus, common on
bread or cheese, magnified 560 times.

partly open under-sink cupboard, and who knows what interesting things it holds. Usually there are only paper towels, but today he could be luckier, and someone could have put something away in the wrong place.

He crawls forward into the enticing dark, a miniature detection machine; eyes wide, and vapor-detection senses—babies on average have more taste buds and smell receptors than their parents—on the alert. He pads into a dense fog of radioactive radon gas that drifted up from the basement stairs overnight. This was stirred upward from the kitchen floor by the family's footsteps, but in here it just swirls around the baby's hands and knees. Some is swallowed. If the house were built on granite rocks there might be enough for lasting danger, but babies have powerful DNA repair systems—especially crucial with their active growth—so virtually all the damage will be repaired before the week is out.

Soaking into the baby's lungs are more substances swirling in the invisible fog. One is the common wood additive hexanal, trickling out from tiny pores in the plywood; another—especially if the family's been adventurous and done some repair work or put in new counters in the last few years—will be wisps of the universal stiffening agent formaldehyde. On a warm day, or when the heat has been kept on overnight, the levels down near the floor will be 100 percent higher than in the main house air. Some bounces harmlessly off the baby's clothes and hands; a little gets suctioned in from the baby's fast-breathing efforts to power itself, and ends up coating its lungs. But this too will soon be dissolved in the universal detoxification chamber known as the liver, as long as levels aren't too high. In a few hours some extra exhaled puffs of carbon dioxide—the formaldehyde's breakdown product—will be the only signs of its presence here.

The formaldehyde that slips past the baby explorer floats upward in the kitchen, but if there's a spider plant perched around, it doesn't stay floating for long. Spider plant shoots have a remarkable fondness for formaldehyde, especially when it is airborne. Microscopic holes open on the backs of their leaves, suctioning formaldehyde from the air. A few days from now, and the plant's roots will seem strangely energized, growing faster than ever, powered by this neatly captured aerial fertilizer.

The baby's in luck. Amid the harmless paper towels is a misplaced bottle of powerful cleaning fluid. These bottles have been as important as vaccines in fighting infectious disease and raising life expectancies: family mortality rates started dropping sharply in nineteenth-century cities, often decades before vaccines were available, wherever there was enough water and bleach and detergent.

Despite the waiting bottles, fungal cities are carving their way into the plastic and wood lining of the cupboard. Regular cleaning would get rid of them, but who bothers to scrub down here as often as he or she should? Fungi will carve their way into virtually any substance that has a toxicity less than that of uranium, which is why this cupboard, the refrigerator's rubber seal, the boy's unwashed socks, and, especially, that nice vulnerable bread lying exposed on the counter, are enticing sites for the digging parties. Kitchen fungi can produce up to one kilometer of fresh microtubing in a day. It's a strange cousinhood to have around us, for although we separated from fungi at an immense distance in evolutionary time, we still share many of the same genetic instructions. Fungal spores that have floated through the upper atmosphere to get to this kitchen—bobbing, unnoticed, beside the passenger windows of speeding jets—are likely to be sun-blackened by the operation of the same melanin chemical that is in our skin. Every kitchen is full of such uncanny parallels. The humble pea plant on the window ledge, for example, oozes out a chemical nearly identical to the hemoglobin in this family's blood, only in the plant its job is to hold oxygen deep in the potting soil.

Many of the arriving fungi can deftly steer around us in the kitchen air, using the flow lines of our airstreams to skim past human obstacles. The gust of the baby pulling open the door gave the fungi a good boost to soar to the food they crave. The furry spots we sometimes see on food are a giveaway sign that one of their colonies has safely taken root, and has grown an interlinked hydraulic feeding network that is such an engineering masterpiece on their tiny level that it extends up to be visible at ours.

The baby pauses to reconnoiter, lured by the pungent ammonia vapors from the misplaced bottle. Ammonia's ability to combine with water during cleaning also lets it combine efficiently with the micropools of water in the odor detectors within our nose. These ammonia vapors that lure the baby are the most abundant industrial chemical manufactured. The process used is remarkably subtle—the factories pull one of the raw ingredients needed directly from the earth's atmosphere—and it was perfected only under the pressures of World War I, by Fritz Haber in Germany, when a British blockade stopped the traditional ammonia-rich imports from South America that the German army needed to make explosives. Without Haber's invention, encapsulated in the humble cleaners today, Germany would have run out of explosives. Haber didn't get much pleasure from his invention: he was reviled by non-Germans as the man who prolonged the war, and even with the Germans he didn't fare too well. His parents had been Jewish, and as an old man the Nazis forced him into exile.

This baby-attracting chemical is also poisonous to drink. The family dog has far better smell detection faculties than any of the humans, as well, generally, as a fondness for the smallest member of the human family; as a result there's a bounding and face-licking and general infant-extracting ruckus as the dog rushes over to rescue its friend from the open cupboard. The baby giggles and tries to wipe his face; the wife shoos the awful, slobbering canine beast away. The dog keeps on wagging its tail though, delighted to be a part of things and stirring more of the low-lying radon in the process. Dogs can smell better than humans, not because their odor detection cells are built better than ours, but simply because they have far more of them scattered inside their wet noses.

Some people don't like dogs, but this is no doubt because they've been improperly introduced, or work in government tax offices. Anyone in a family can express his or her real feelings about someone else and get away with it through the simple expedient of looking the dog in its eyes and patting its fur and saying what he or she wants while pointedly *not* facing the family member who is really being addressed. There's also the pleasure of having this comprehending beast huffing and slobbering and nodding to you in deep commiseration during the soul-unburdening talk. Dogs are excellent for a family's health. The stimulus from patting dog fur seems to massage our brains. A good bout of fur petting brings together nerve signals that are rarely linked: there's the firing of the surface-touch receptors in our fingertips, and the warm-temperature detectors slightly deeper down, and then, in those moments of full-out they-don't-understand-me-either skin-pummeling rubs, the sort which domesticated canines put up with, as they're usually polite enough to give only the slightest squeaks of distress, the petting even gets our deep level bulbous-shaped pressure receptors, the ones normally untouched far beneath our fingertips, to fire their impulses brainward, too.

It works so well that regular petting of a dog almost always lowers your blood pressure; it even, as pleased researchers at the University of Pennsylvania have confirmed, raises the survival rate for heart attack patients. About 35 percent of the patients without a pet didn't survive for the length of the study; only 6 percent of the patients who had a pet died. Dogs can also be instructional. Mark Twain always gave his pets names such as Apollinaris or Zoroaster "to practice the children in large and difficult styles of pronunciation." Under the circumstances the enormous cost of dog food—it generally comes to 6 percent of an owner's total grocery bill—makes a little more sense. Despite all their merits, however, dogs are no longer Americans' most popular pet. With work demanding ever more time, during the mid-1980s cats took over.

It's true some cultures appreciate these tail-wagging health aids a little more than we consider entirely proper. In France there are twice as many dogs and cats as there are children, and wealthy women will often bring a dog to a restaurant and let it eat from the table. In parts of Polynesia dogs could have been brought to a restaurant also, but wealthy French women might have been upset by what happened next. Wherever a creature filled an island's niche of being one of the larger sources of mobile protein around, it made eminent ecological sense to use it for meat.

Up goes the baby to the counter once more, close to the intriguing radio/CD player and a safe distance from the pancake batter. The dad grabs the frying pan, puts in thick dabs of butter—for hasn't he read something distressing about margarine?—then adds a little extra water to the pancake mix. Dissolved lead might come out when the water faucet starts to run, having diffused in the pipes overnight from their soldered joints. If you let the faucet run for a few moments before filling any container, you'll avoid any such lead, but other chemicals are likely to come pouring out anyway, which is one cause of the Great Sperm Disappearance.

Something has been happening to men's sperm in this century, and nobody knows exactly why. College students of 1929 had 90 million sperm cells in each milliliter of semen, while their successors in 1979 could manage only 60 million, and today it's even lower. One likely cause stems from the way

Two epochs of recording technology. *Top:* a compact disc, showing the digital notches under the protective plastic surface. *Bottom:* an old-fashioned long-playing record. The more the stylus wobbles through the plastic canyons, the louder the music.

farmers often feed or inject their cattle with female hormones to get the cattle to put on weight. These hormones are chemically near-identical to human ones, and they stick in food and run off into drinking water. At one point London's drinking water from the river Thames was so laced with female hormones that a cupful would give a positive reading on a sensitive pregnancy test. Other chemicals can also be part of the trouble—we encountered the hormone-resembling ones on the plastic wrap earlier. Expensive bottled water is not always an improvement, as labels declaring their products to be natural or pure or bottled at the source can still cover a multitude of origins. Sometimes bottled water will be what you expect, water from isolated rural aquifers, untouched since the Stone Age; sometimes though it's just from ordinary wells, located near huge factories that pump in the natural-seeming carbon dioxide. In the case of one ingenious British supplier, the majestically pure water was produced by holding empty bottles under an ordinary faucet then sticking a glorious mountain scene on them. Whatever the source, this morning's drink of orange juice can only help: the vitamin C it contains aids sperm in its defenses against chemical attack.

The baby watches his dad in awe and even the ten year old looks up from his computer game. When there are naked flames, even the mildest of men can feel the primeval urge of the blacksmith pump through them.

It is Dad the Fire Wielder.

The Barbarian Swordsmith.

The squeezer of the teensy little button that ignites the safe gas burner.

The gas flame ignites, and molecules within the metal frying pan begin to vibrate faster from the rush of miniature explosions underneath. Some actually lift off into the food (though even if it's an aluminum pan there are not enough molecules, however long the cooking, to accumulate in the brain). Water droplets within the butter chunks start vibrating with the heat and themselves explode, and dioxin residues move up to the melting edge of the butter. Cows are the perfect dioxin-collection machines, chewing through great amounts of grass or grain that's constantly coated with fine traces of air-deposited dioxins. The butter pat bounces from the unleashed energy; in swirls the torrent of pancake batter, sending even more spattered fragments of butter flying up. At first it's the delectable butanedione liquid, which gives butter its flavor, but then harsher carbon granules soar up in their millions, floating like sharp-edged meteorites. The air fills with this abrasive rubble.

The heated butter could be dangerous, but the family's blink mechanisms simply switch to a higher gear. Normally we blink twenty-four times per minute, less if driving or happy or reading, more if angry or talking to a stranger. Children blink less than their parents, and cats, near catatonia at the best of times, make do with just two blinks a minute. The baby is closest to the line of meteorite impact so it responds first. Dangling blankets of skin weighing one-fiftieth of an ounce—the weight of a hummingbird's tongue—slither down over the vulnerable eyes. If our eyelids simply ground the microrocks into the eye surface the action wouldn't be especially useful. But with each blink extra salt water and flotation agents are injected from the tear ducts. The muscular skin blanket also curves itself in a gymnastic part-curl, not closing down all at once, like a fleshy guillotine crashing because of gravity, but rather rippling in sequence from the outside in, like a sea creature fluttering its tendrils in a sinuous wave. (You can just about see this directional twist if you are impolite enough to stand face-to-face with a fellow blinker and stare; you might as well, while you're at it, observe the other feature, that our more slender lower lid always rises first.) The push from the outside in rolls the landed rubble toward the small pink collection area at the edge of the eye nearest the nose. An exit tunnel there is quiveringly open, and leads down to the immense cavern of the nose below. The microboulders are shepherded to the pink area, a final blink squashes them through, and they'll only emerge several minutes later, free-falling or stickily dangling from the bottom of that tunnel, far from harm's way.

People tend to synchronize their blinking rates with each other, so the baby perched on the counter serves as an early warning system: the rest of the family, watching him from back at the table, is likely to start defensive blinking even before the aerial fragments reach the table.

High above, screwed tight to the ceiling, the electric circuits of the family's smoke detector begin trembling excitedly as the spatters float all the way up. There's a live radioactive chunk inside, which for years now has been fruitlessly beeping out alpha rays in order to create a smoke-sensitive aerial beam. Once such radioactive materials were dominant in open, intragalactic space and not humiliatingly consigned to life stuck inside a tiny plastic box. But now the radioactive chunk hints at its once-mighty power, as the alarm blasts away.

The dad is now entirely carried away with his fire-wielding force, and doesn't turn the flame down, so smoke rises into the enclosed room air. Under extreme conditions we can see the tiny burned butter granules rise as what seems to be a dark film of wispy smoke. The wife nods to her ten-year-old—a teenage daughter is long past the age of obedience—who steps to the sliding glass door and is about to open it to the back lawn.

The wife thinks the outside air will be entirely refreshing, but this is questionable and depends on what you mean by fresh. There are very old atoms waiting there, many having floated or stuck on our planet for several billion years, clustering up against the glass window. Each atom is a tiny speck shaped a little like a spinning mini solar system, tumbling weightlessly outside. The marvel is that we burn the stuff to breathe, but this in truth only applies to the waiting oxygen. There's the great bulk of seemingly useless nitrogen and other natural molecules. There are also, depending on location, all the manmade extras.

In Los Angeles what's humorously called fresh air is loaded, in addition to the well-known car fumes, with stupendous numbers of floating grease balls, small enough at their microscopic size to stay up for hours. They're generated by the estimated 3,000 tons of meat fried at fast food or other outlets in the area, and researchers have found that they compose 4 percent by weight of the fine-particle air pollution in the metropolitan area—a total of 46,000 pounds of flying burger bits per day. Blinking will push a lot of it across the eyes and down through the lachrymal duct to the nose, but a lot is breathed in as Los Angeles residents give their tracheal linings a burger-rich speckling.

The boy tugs and the sliding patio door is open, stretching wide. All the gases generated in the chemical warehouse of the home this morning—what had made the room bulge out balloonlike, as these particles struggled to get loose—finally have a chance to escape. There's a whooshing gush, as out it all pours, the house's unique pollution signature: the pancake particles; under sink hexanal; nitrogen compounds from the stove; and slow-bursting radon from the basement. There's even likely to be some CFCs slipping out from the fungus-weakened sealants at the back of the refrigerator. The smoke

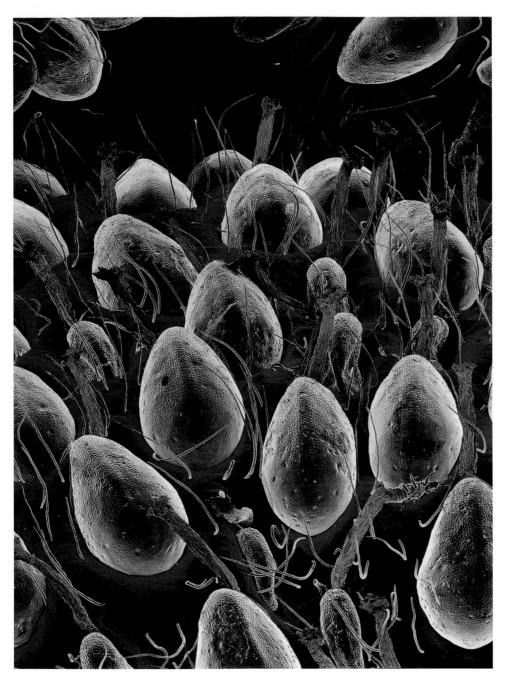

The swollen ripeness of a strawberry;
the green ovals are the seeds.

alarm quiets as everything speeds away across the patio and lawn. Some of the nitrogen compounds will travel entirely around the world in their twenty-year life spans before they decay. The CFCs will float up to the stratosphere, with the strongest-surviving of them bobbing in position for up to an estimated 460 years—a bequest from this morning's family to high-altitude explorers, if there are any, in the middle of the twenty-fifth century. But the family notices none of this, for everyone's at the table: the pancakes and maple syrup and toast with all its toppings, the orange juice and milk and fresh-brewed coffee, too, ready to be consumed.

2

breakfast continues

From the buzzer perched above the distant front door, a tidal wave of air suddenly begins to push into the house. A little bit races along the stairwell to the empty upstairs; the main wave pushes straight ahead, to the kitchen just across the front hall from the door. Some swirls over the fungal cities under the sink, but most of the crashing wave froths straight into the feeding family. Many of the high-speed shock waves crash against their bodies, bouncing off to eddy uselessly in the kitchen air; other fragments find the winding tunnels of the ear that lead deep inside each head.

And nothing at all happens.

The tiny metal clapper that banged into the bell pulls back and, still powered by the hand pressing on the doorbell outside, still a part of the first ring, now—about two milliseconds after the first strike—moves forward again and gives a second great resounding whack to the curved metal bell. Another tidal wave of air leaps out and races into the kitchen and hits the humans.

And still nothing happens.

The fiery sky of a sodium compound magnified. Sodium is omnipresent in the home, from a doorbell battery to the human nerve cells that hear the bell.

The metal clapper hits again, for the continuous buzzing ring which we hear is actually made up of numerous repeated slams, and there's a third house-filling air blast, a fourth, and a fifth—air waves blasting into the human ear tubes each time—and only at the 120th repetition, still a part of the first buzzing ring, a quarter of a second into the barrage and with the fungal cities thoroughly stirred up, the raised energy of the rushing air burrowing into chairs and curtains and fork-pierced raised pancake segments; only then does the first of the humans begin to react. For we live in an entirely different time scale from the mechanical objects around us; our reflex networks work through wet nerves and sopping brains, and we can't hear the separate strikes which made up such seemingly continuous sounds.

Slowly, with the gravest of deliberations, the father's eyes start rotating toward the sound source, sloshing in the lubricated eye sockets of his skull as six miniature muscle cables hooked into his socket bones start to pull. About twenty-five more aerial shock waves batter into the kitchen, and then, with their slightly slower reflexes, the wife and son begin the eye rotation, too. A moment later the baby's eyes start to respond, but by now the father, the comparative high-speed marvel, is gearing up to an actual body twist. Soon all the family's tendons are cranking and creaking, as they pull on their skeletons in order to get their heads and then their whole upper limbs to follow. In under half a second—a mere 250 repetitions of bell-thumped air waves from the first sound pounding in—the operation is done: all heads are tilted toward the source. Data streams of location, time, vehicle noise, and even door knocker banging can finally enter conscious awareness.

Although the father is likely to have started to turn first, the wife still has a chance of getting the hallway and the door beyond into view before him. The reason is that women have more receptors away from the center of their retinas than men do and so invariably register better on tests of peripheral vision. Men usually have to get a more face-on view to detect things as well. They also have to struggle in their responses to virtually every other low-level input. Almost every wife and daughter can out-see her husband or brother in low light. She can also usually hear high-frequency sounds better—such as any squeaking high-level harmonics from the buzzer, which is the cause of the dog's distressed barking—and, as we'll see, can usually detect dilute tastes better.

If there were a cat around it would do even better at peripheral vision. This is because cats have more rounded eyes than humans. These are great for bringing in the big chunks of light cats need for dusktime hunting, but the shape makes it difficult for the center of the retina to get a sharp focus— whence the curious way cats gaze into space with seeming mystic insight all the time. It's to make up for their focusing problem that they've developed enhanced peripheral sensors.

Ultrasonic shrieks, too high even for the dog to hear, are leaping out from the quartz crystal in the wristwatch, which the wife rotates upward to glance at. High-speed electrons lift off, racing upward from the luminescent coating on the dial surface, but these smash futilely against the thick glass dome, and stay trapped within. (Watches from before the 1960s used painted radium, which sent powerful gamma rays splintering through the dial case.) The habit of wearing chronological devices on the wrist largely dates from World War I, when it became stylish to copy Sopwith aviators who didn't have the time to fumble with an inside pocket. Dividing a circle into sixty degrees or minutes had been a standard practice since ancient Babylonian times.

After the turn, the next thing a startled family will likely do is get ready to swallow, for no one likes getting caught with his mouth full. Chewing speeds quickly rise from the stately molar descent of 0.08 miles per hour of ordinary times to a wind-rushing 0.12 mph now. Face muscles distort, and the motor nucleus in the brain controlling them is suddenly burning more glucose than before. Women have better control of the numerous tongue muscles and their direct lines to the brain than men, and so almost always finish chewing in fewer bites, as one can prove by surreptitiously counting. There's no reason to engage in the strange habit of chewing each mouthful thirty-two times. The practice, called Fletcherizing, after the turn of the century crank Dr. Fletcher, has no better justification than the fact that we all have thirty-two teeth. Children have more, for their future teeth are already in place in miniature form below their current ones, as X rays show. The tongue that pushes all the rapidly ground food into place reaches its maximum weight at age twenty-two, after which a small but steady decline begins.

The surprise at someone coming unannounced to the house on a Saturday morning is a sign of how much we've become used to being isolated with our families. Go back enough generations, and almost everywhere you'd have seen people rambling in all the time. One Nuremberg family of the mid-1500s reported, on a single Saturday, visits by peddlers, a jester, ice merchants, the priest, knife sharpeners, various servants' friends, a glass repairman, neighbors, a baker's apprentice, neighborhood children, and—the children's favorite—the leech carrier, offering hygienic or purely recreational bleedings. Hardly anyone in previous societies was granted much privacy—even Louis XIV was assumed to not mind his courtiers standing around as he urinated in the Versailles hallways—so there is no model for families being locked away all day. It's a bit extreme for our taste, but better than Mao's China. For many years most families there were allowed no visitors at all—not even relatives or close friends—unless the family had permission from a local block committee.

Once the hurried swallows are done, air is sucked in for a quick puzzled discussion about who it might be. The dad gets up to see, but even as he starts walking one inner part of the swallowing is still not completed. Liquids tumble in the free fall of gravity all the way down, but solid food goes slower, at the rate of the peristaltic waves of the esophageal tube pushing it along. There are about eight seconds—the rushed swallow long forgotten and the dad well on his way from the table—before the lower esophageal sphincter finishes widening, and, in a shared sequence of silent plops, four balls of the family's food join the earlier orange juice and toast, slapping into four waiting stomachs.

The baby bounces with arms raised skyward from his high chair, not wanting to be left out of the excitement, but the father isn't going to take him. The dog's efforts at bounding along are less thoroughly resisted though—who knows for sure what's out there? So one tail-thumping specimen of *Canis familiaris*, the rest of the family left protectively behind, accompanies Dad as he heads for the door now, pushing and wading through the kilograms of air.

There's barking and eager if not particularly accurate jumping, as the gush of *enemy?* vapors—undetected by the human, but frantically exciting to the canine—pour in from under the door. The dog is motioned away though, and the doorknob cautiously rotated. A grinding of metal parts now begins within the door mechanism as parts of the doorknob melt onto the rod that carries the turn to the latch. The pressure from an ordinary human forearm can do this because only scattered, isolated regions of the doorknob mechanism are melting—it happens only where fragments of metal are a little higher than everywhere else. But those scattered actions are crucial, for it's the friction they now generate that turns the latch. As you continue turning the doorknob, you quickly tear apart the first group of welded sections and replace it with another. (Polishing silver works on the same principle: for brief periods the point of contact between the cloth and the silver gets so hot that the silver flows over impurities and shines.)

A final shush to the dog, and the home owner steps out, to be blasted—this feels really good after you've been inside all morning—with energy from an exploding hydrogen bomb floating overhead. These arriving photons entered the upper atmosphere just fractions of a second before, when his hand was already off the metal knob; they were speeding through cold space in the orbit of Venus two minutes and twenty seconds ago, and only left their source—the surface of our Sun, of course—a brief six minutes earlier. Their history before that was more intricate, if a little slower: the originating photons began their journey up from the central depths of the Sun 10 million years ago, when dinosaurs were long gone, but *homo sapiens* were still in the future. A lot of these photons thunking down end up snugly caught in a cotton shirt or jeans, their 93-million-mile flight cut short just fractions of an inch from the skin below. But some make it farther, and reach bare cheeks or forearms.

Why do these fragments of a distant star feel so good? Their ultraviolet energy makes our endorphin levels rise pretty quickly. Endorphins resemble opium closely enough so that they trigger our brains' pleasure receptors, deliciously but safely echoing the effects of this drug. Other fragments of the light work on our eyes to trigger pathways leading not just to the usual optical processing centers of our brain but also to the pineal gland, where the melatonin chemical is made. In extreme cases sunlight can adjust the amount of melatonin leaking out and cure the depression some people get in the low light of winter; possibly in ordinary cases this has some satisfying effect, too.

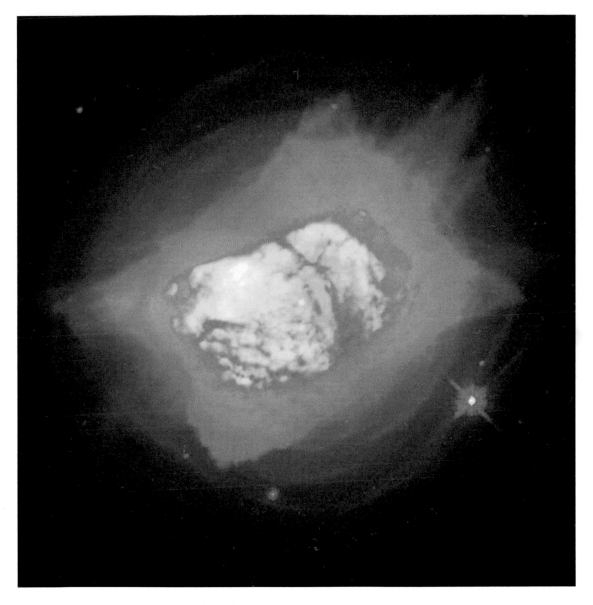

This star exploded 3,000 years ago, but its radiation is still reaching us, shattering DNA strands that our bodies must constantly repair.

Finally though, the light is also good for our health. Steroid molecules are packed in the top of our skin just waiting for the light. The impacting energy flicks those molecules into a new shape—what we know as vitamin D—and with that change they float off into the bloodstream, to reach the bones and other regions for healthful effects later. Stretch your arm out to get more of the rays, and your solar-powered vitamin factory switches to high.

Since the northern latitudes suffer from weather that ranges between miserable and very miserable, the effect is especially important there. That's why people who've had the misfortune to live for generations under those cloudy skies, such as the Scandinavians, get especially excited about sun exposure and come equipped with the most efficient reception units: pale, photon-transparent skin. Major suntanning is a different matter. The twentieth century began and ended with educated people fearing it. Only for fifty years or so, starting in the 1920s and inspired by Coco Chanel and other adventurous visitors to the south of France, was a contrary view widely held.

The dad stops to take a good sniff of the fresh air; proud of his manicured lawn and the shrubs and roses in front of the house. The house bricks behind him are a carryover from standards set centuries ago. They're just big enough for an ordinary-size mason to lift with his left hand: the right would be busy using a trowel. Some of the sweet vapor in the air is a desperate chemical message screaming out from the shrubs as they send an odoriferous alcohol upward to spotlight the point where they're being attacked by caterpillars or other gnawing insects. On a summer's day this SOS signal reaches any wasps lazily circling overhead, and leads them vectoring down to tear into the unsuspecting caterpillar. From the sensitive rosebush, another gas is being released to counter awful mass-fungus attacks. This gas leapfrogs over the fungus and signals the further reaches of the rose's leaves to pump food-blocking barriers around the fungus, starving them out. Even the attractive smell of freshly rained-on ground that we sometimes notice has a similarly desperate source: it's produced, in large part, by soil-dwelling bacteria spraying out poisonous blasts to keep away competitors. If someone's been too exuberant and planted the flowers too closely together, their roots are likely to start streaming hydrogen cyanide into the black soil to annihilate their neighbors, but occasional trickles will break through the surface, luckily at levels low enough not to disturb us.

The trees up above are a little more harmonious than the shrubs and flower beds. There's some evidence that they're hooked up by invisible gas communication lines, using puffs of ethylene gas to warn each other if one of them is coming under any caterpillar or fungal assault. The trees pump defensive chemicals into areas under attack, and where it's really bad, send out additional chemical puffs to help cut off a damaged leaf and send it, plus its bewildered caterpillar, circling in free-fall to the ground. Any caterpillar that doesn't survive the journey will be partially ingested by tiny tubes the tree sticks up from its topmost roots, in a sensible ecological cycling.

The dog bounds out, but the deliverer doesn't have to be scared: the creature's just racing to the trees on the lawn's edge, where the nose sniffer's heaven exists. Exuberant tail wagging helps this, by fanning odor communication molecules more briskly into the air. Any earthworm will detect the dog's shadow racing by, and will send danger chemicals back down the tunnels the worm emerged from, alerting other earthworms to stay down. You can usually trust that a new dog you meet in the neighborhood will be safe if its head is up or level with its body, if its tail is loose, and its ears relaxed. The time to climb a tree is when the head is bent down, the tail's taut, and the ears are erect. (If it's also crouching and staring at you, it's wise to climb quickly.)

To raise a dog that will let deliverers survive without blood transfusions it helps to have the dog handled by several different members of your family when it is very young—trainers have found this one of the best predictors of having a mild adult dog. It also helps if it's the right breed. Beagles and Labradors are on one side of the spectrum; on the other is the breed that was created in what might have been the plot of a particularly low-quality horror movie. Around 1900, a German dog catcher decided to save a select few of the dogs he collected from being killed. He chose the ones that had "the most vicious disposition, and the worst temper." The products of these experiments were crossbred with the next generation of most vicious strays. It took strenuous effort to make sure there was no backsliding over the generations, but with continued selection for vicious temper the procedure finally worked, and Herr Dobermann had the honor of getting a new breed named after him (although it lost the final *n* somewhere). In fairness to Doberman lovers, some branches of the modern breed have been crossed with terriers and greyhounds to make them less like their creator.

The deliverer's back at the sidewalk to check the house address against his clipboard. In Tokyo this would be especially difficult as many streets have no names. Each block may be named, for example, after a local shrine, but even so, the house numbers on it won't necessarily be in boring sequential order. Until about two centuries ago houses in Europe and America had no numbers. But when governments needed a better way to keep track of their citizens these useful tracking numbers appeared.

The deliverer climbs back up the steps now, the nearest ants turning curiously; the impact of his footsteps causing a great resounding thud on their incubation chambers below. He has a package in one hand, and now, in front of the dad, brings the clipboard up to be signed. But can this be done in just any posture? Social psychologists have noticed a shuffling almost as complex as the way ants frisk each other with their antenna: human males seem to have signed an agreement that they will never stand in direct proximity if they can avoid it. Instead, they take up a strangely distanced, now-don't-you-take-offense posture, with their chests not facing the other's directly. Such positioning would identify them as males even in the faintest sighting from down the road. Women, oddly unterrified, are generally willing to face each other directly, and stand closer than men will. They also give the little supportive nods and grunts we like from close friends during a conversation, and will regularly look at each other's eyes as they talk. Men, and especially men who don't know each other well, don't. If by some unfortunate chance their glances do cross, so that they're looking directly into each other's eyes, they almost joltingly flick their gazes away. The electrical conductivity of their skin goes up during conversation, and stops only when they finally separate.

It seems silly when replayed in slow motion on a video, but at least two individuals who know this code can usually get on all right. It's when different sexes meet that things can get confusing. Because men are so conditioned to sidle up from the side, thereby showing nonhostility, but women prefer to be approached cleanly from the front so they can see the other person, a man who approaches sideways in an effort to put a female deliverer at ease is likely to be met with suspicion.

The father has to sign for the package. The muscles that control his signature are simple to operate, but the inertia from the massive swinging hand sends the pen jerkily oscillating at the end of each stroke. This is why it's hard to forge someone's signature.

A pencil. Note the spongy wood—pencils are often made from trees 150 to 200 years old, to make sharpening easy. To make a pencil, the chunks of clay and graphite we call "lead" are placed in open halves of wood. The seams where the halves are sealed are often visible under the paint.

The dad can barely see what's going on around him now, for when we concentrate hard our visual processing ability fades away on the periphery. Our brain waves also change, briefly rising to a frantic cycling rate more than twice what they are normally. The blanking out of vision is especially strong in the first fifth of a second after taking in a new bit of information, such as a request to print one's name with each letter capitalized and in a separate box. Sounds are especially hard to take in then, and if someone speaks to you precisely at that moment, you will only dimly recognize the voice. Family members generally recognize this, and subconsciously adjust their speech rhythms so as not to interrupt during these brief phase-outs. Outsiders are not always so kind.

It should be no surprise which family name is most likely to go on the form: it's Smith. There are an awful lot of Smiths. It's the most common name in America, by far. The stands of every NBA basketball court could be filled exclusively with Smiths, you could have every team member be a Smith, employ as commentators and cameramen only people named Smith, have every actor in the commercials and every sideline cheerleader be selected from Smiths, and there would still be vast numbers of Smiths left at home to watch the whole thing. There are dozens of them at every big concert hall; the odds are better than even that there will be at least one in every large airplane overhead. There are over 2 million people named Smith in America according to Social Security estimates. With the equivalents in other countries—Lefevre in France, Kuznetzov in Russia, Kovacs in Hungary—the world total is probably around 5 million.

Many of the names on the clipboard—the Coopers and Bakers and Thatchers—make an interesting snapshot of the jobs that were common near the end of Europe's pre-industrial period. The reason is that this is when many family names were first consolidated, as with addresses, often for the same reasons. Yet not all the names that survive to be signed with a flourish today are quite so insightful. With stunning lack of originality, administrators in Greece and Ireland and Scotland all seemed to latch onto the habit of labeling as many people as possible with the names of their fathers. If that name were Michael, then Michaeledes or O'Michael or McMichael was the result. You can't do much late-medieval genealogy with that. And then there are the last names that might be illuminating but are still a little embarrassing for modern families to delve into. Kellogg sounds distinguished and hygienic enough, but similar names derive from "kill hog," commemorating a family of pig slaughterers. The common "ski" ending in Poland, apparently, comes from the period when most people there were little more than slaves. If you lived on the estate of say, a Count Pilsud, the easiest thing for the overseers to do at stocktaking time was give you, and everyone else slogging away there, the inventory label of Pilsudski. If the authentic Count also came to be known by that extended name, distant bearers of it could believe, or at least claim, that they were descendants of one of the elect.

• • •

The dog's ears lift up as it hears the sound of something familiar approaching around the corner. It starts barking in a different way, alerting the son, who knows what this means. He arrives outside just as a car pulls up, and his best friend, holding a white cardboard box, climbs out. Breeds such as Labradors hear sounds $1/8$ of a tone apart, which turns the generic rumble of a car into a crisp, unique chord. At one time the friend could have arrived on the fragile spinning gyroscopes known as the bicycle, but he's never been the leanest of individuals, and with distances these days, plus who knows about crime, that's gone. In England an estimated 30 percent of thirteen-year-olds have never been out of an adult's sight in their lives. In America, the figure's probably lower, but on the way up. The driver of the car, as chubby as the progeny he's dropped off, nods awkwardly to the dad on the porch, acknowledging that the driver should be embarrassed at off-loading his kid for the day without prior arrangement, but not so embarrassed that it stops him from gunning his engine and driving off to freedom.

Blood pressure and heart rates go up with all the excitement, but it's unlikely to have the same effect in the son as it has in his friend. About 20 percent of families carry a gene that makes their levels of the blood-clotting agent fibrinogen easily rise when they're excited. This pulls more cholesterol to the inner walls of their arteries. If the father and son have this gene they'll start to get a miniclump of oily cholesterol buildup now; the friend, even if overweight, is safe. Families with this common flawed gene will also get fibrinogen rises when the air suddenly gets cold, which is one reason shoveling snow in cold weather is so dangerous for a segment of the population. The worst possible time to shovel snow is Monday morning, for that's when heart attacks are at their weekly peak anyway. Saturdays are the safest time in the week, probably because general stress-hormone levels are down.

Exhaust fumes roll in from the accelerating car, but the humans on the porch don't get swamped, for the lawn's defenses get to work. The trees on the sidewalk's edge widen the open tunnels leading into their leaves, and minia-ture pumps inside start pulling in the car-released gases as they hurry by. Even a small oak can have 40,000 chemical-defending leaves dangling from its branches. The result is that dozens of pounds of poisonous gases and even the occasional toxic metal can be suctioned in each year, then shunted down to the hydraulic roaring roots to be deposited, as drops of poisonous liquid, far from trouble, deep below. The more defending trees you have, the greater the aerial scooping protection. The porch-gathered humans are now engaged in the footwork and mix of grunts of greeting and farewelling hand lifts needed to send off the deliverer and replace him with the visiting friend. Earthworms grinding through the soil absorb some of the deposited pollution, without too much damage; even tinier microlife in the stacked clay layers around the roots will get to work in the next hours and weeks crunchingly detoxifying much of the rest.

Air fumes that make it through the guarding tree barrier are next attacked by the lawn. Mists of invisible moisture are rising from the grass. This moisture coats the smallest pollution particles and makes them stick together and so suffer the worst possible fate for airborne travelers: a sudden overbalance of excess weight, and a frantic tumble down to the waiting grass below.

The *hungry* grass.

Tiny pores open wide on these miniature plants, too, pulling in the pollutants just as the trees do. Signals transmit from the grass's roots, alerting the soil's microlife that there's a food shipment coming down. But grass blades are so much smaller than the first line of defense at the trees and they're equipped with so much less shunting and detoxifying mechanisms, that at times individual plants are overloaded. Excess pollutants are dumped down to the roots, especially if the delivery van has been waiting, exhaust pouring, with a poorly tuned engine that's just spraying out the oxide clouds. It might look like the grass blades are just isolated towers of chlorophyll, each standing alone, but the lawn outside our house is an immensely busy ecosystem, with the seemingly isolated blades actually thickly cabled together underneath the surface. Leathery fungus arms stretch in a thick network connecting each part with others, and where one section gets damaged from absorbing the van gushing pollution, another section, safe from damage farther away, lush with excess nutrients, now starts to feed it. Attach a tiny microphone to one of those cables this morning and a miraculous gurgling sound would be heard as water and sugar and vitamins and amino acids all come pumping in.

Gutenberg's legacy: the corner of a computer chip, enlarged 60 times, active in appliances from modems to microwaves.

Back inside, the baby bouncing unhappily in his high chair, distressed at being left out, the chubby friend politely greets the boy's mother, then passes her—this is a regular ritual—the cardboard box. Everyone peeks inside, except for the teenage daughter, who is still ostentatiously sitting apart, but she wouldn't join in now if someone paid her. She finds it just unbelievable that they're willing to allow more food in this kitchen. Because they are Danish pastries though, everyone else agrees that the pancakes should be put, at least temporarily, on hold. The hand-delivered letter the dad has received is put aside on the table, too. The microwave door is opened, the cardboard box of pastries pushed inside, the timer is set, and the start button is pressed.

Is this wise? The theory is that all the microwaves are supposed to stay inside there, but crumbs or grease stains building up on the hinges make the seal less than perfect; a baby who's been eager for close-up investigation—with a sharply tapped spoon on the fragile protective grille over the door—will make it worse. A few of the high-speed microwaves almost always spurt out, squirming in five-inch-long shimmies as they explode into this family assembled raptly before its machine. Usually, regulators say, it's not enough for anyone to notice, but try telling that to Peter Backus, an astronomer with the international radio telescope in Parkes, Australia. When he first detected inexplicable signals at 2.45 megahertz on his viewing panel—and, even better, realized that they were coming at the same time every evening, precisely when one particularly suggestive stellar formation was appearing in the sky overhead—he was convinced he'd achieved the astronomer's Holy Grail, and detected alien life. Alas for the world's headline writers, Backus was a cautious sort and did some more localized checking first. As his shift started, the previous shift was ending. It always took a little while for his colleagues to get settled, which is why he usually had just enough time to switch on his radio telescope registers . . . and detect them heating up their frozen dinners in the staff microwave—set at the usual power rating, which is 2.45 megahertz—downstairs.

The friend, watching as intently as the family, is gushing tremendous numbers of alien bacteria into the kitchen's air. Humans pump-wheeze dozens of gallons of bacteria-rich breath clouds out each day, which is several quarts in just a brief wait. In the confined space of a breakfast room or kitchen everyone's going to get it. These bacteria are tough creatures that first evolved several billion years ago—almost at the first moment when the earth's surface temperature cooled enough for stable life—and have survived, truly dominating all other life forms on the planet, ever since.

What they didn't have to reckon with, however, was us.

Almost all these live microcreatures in the clouds the chubby friend is spewing out have a tremendously short life span here. The chlorine-rinsed sink counter and the bare metal stove top and the sleek plastic light switch are entirely inhospitable to their life requirements. They die by the thousands as they gently touch down, these parachutists from an ancient war. Only if you leave such nutrient-rich culprits as a slightly damp cloth or sponge around—both of which are dense in much-sought water, and with nourishing microfood fragments scattered within—will randomly landed bacteria find a safe refuge on which to grow. This is why most families end up with their own distinctive bacterial populations in their kitchens—and why eating over at a friend's can leave you with a low-grade upset stomach to which your hosts, after these years of exposure, are totally immune. Some people try to fix things up by using bleach or other disinfectants when they clean, but usually they think the right thing is to pour it down the sink. This is a manufacturer's dream—you get to sell something that people buy and then immediately pour down the drain so they have to buy more of it from you! Wiping bleach on the surfaces so that the invading bacteria have to try to survive a soaking in that is a better technique.

The wife breathes in, helping the baby turn over the envelope, and hundreds more bacteria are suction pulled entirely out of the room air, seemingly headed straight down to her lungs. That might seem better for the bacteria, yet the nice quivery lung tissue, all pink and vulnerable just a moment's flight inside, is not so easily reached. Everyone else in the kitchen is similarly breathing in whirlpooling bacterial emissaries from the newcomer. But a family that has defended itself against aluminum, frying pan rubble, and ultraviolet photons, is not going to be bothered by a measly fleet of ancient bacteria. The clouds of bacteria-rich air enter, but by the time the air reaches our lungs it is almost entirely sterile. Most of the bacteria are absorbed by the breathing tubes' sticky lining on the way down.

Other bacteria are carried toward the back of the throat, where with an unnoticed swallow or glug some two or three days hence, they will be heaved over the edge, to be sent down into the streaming acid rivulets of the stomach, from which there is no escape. We never notice when the sticky "escalators" carrying the debris up from the lungs works properly, and I suppose most of us would prefer not even to know of their existence. But for anyone with cystic fibrosis, where these sticky linings clog up, their proper operation is life itself.

Any ten-year-old who hasn't brushed his teeth recently is likely to have brought a rich additional store of bacteria, as an irregularly cleaned mouth is an ideal moisture- and food-laden incubation chamber for certain other species of bacteria. In particular, hunkering where the teeth erupt from the gums, there are likely to be awful spirochaete bacteria, with quick-spinning bodies, writhing like frantic miniworms in the hundreds of thousands all along tooth edges. Flossing would destroy them, for these are direct descendants from the earth's earliest life forms, and die when exposed to oxygen. Dental floss is a huge tugged rope on their size scale, and cracks open the oxygen-barring crust above them.

Even the spirochaetes are as nothing compared with one item that can be alive in the mouth of a boy who doesn't brush regularly. There will be enough micromeat to keep a tiny population of predators on the move. These are the *T. rexes* of the peaceful breakfast smiles: the awful *Entamoeba gingivalis*—a huge quivering thing, many times larger than the bacteria it hunts on the gums. A week or so of enforced flossing would kill the ones in place, but unfortunately that's unlikely to mean they're gone forever. An estimated 50 percent of household dogs have these same quivering monsters hunting on their teeth, and any eager bout of slobbery kissing by the boy will spread the *Entamoeba* right back. Chairman Mao *never* brushed his teeth, ac-

Beauty in a human mouth: undulating pressure ridges on a molar tooth.

cording to his personal doctor, but just rinsed his mouth occasionally with tea. Over the years his teeth became coated with a thick green film, no doubt rich with these microscopic hunters.

The boy notices his sister's aloofness, and with the typical sensitivity of a ten year old, glances to see his parents are occupied, then quickly makes a face and sticks out his tongue at her. Anyone watching with a quick-focusing microscope would now briefly get a glimpse of the great wobbling *Entamoeba* propelled through the air, straight toward her. But if she's lucky enough to harrumph in disdain at his immaturity, that will blow enough air currents to halt the invading monster before it can board, and send it, at its great weight, arching downward, to plummet harmlessly on the barren surface of the table.

While the rest of the family waits for the pastry to heat, the letter that's been delivered is distractedly opened, the perforated tab along its side scattering microclouds of cardboard rubble. Many will make it to the family's nasal air caverns, to join the squirming bacteria already stuck on the lining walkways. There's a bit of peering, even the youngest member stretching over to see, then a final deft flick of the tab, and the item inside is revealed.

Junk mail.

The parents groan and turn back to watching the microwave; even the baby, seeing their dismissal, tries to squeak out a matching groan in imitation. But to the ten-year-old boy this isn't junk mail: this is something addressed to the family c/o his dad, and that means them all, which includes him, and—who knows?—maybe it's important. The envelope announces a VALUABLE PRIZE to be won, so he and his friend start dragging their graphite-leaking pencils over the forms to fill out. The pencils' wood is liable to be much older than the boys, and probably older even than this house: to make pencils easy to sharpen, it's common to use wood from trees 150 to 200 years old.

The actual text and glossy pictures that fall from the envelope are left behind, ignored by everyone in the family. This is a shame, for the analyses worked out by direct mailers—the little electronic shadows of corporate information on us—can be so accurate as to surpass what a couple acknowledges about themselves. In many Protestant families, especially in the South, market analysts found that it's husbands who write the checks or sign the credit forms for buying the family car, despite what the women assert about being involved. The car mailings they receive, accordingly, show the men talking to the car salesman. Jewish couples turn out to be more likely to share car buying, so the car brochures that are mailed to Jewish neighborhoods or families with Jewish-sounding names have a better chance of including a paragraph on what fun it is to visit a showroom together. International variations also exist. In America, women usually purchase kitchen goods, however much their men try to proclaim that they care, they really do, about what goes on in the family kitchen. One big multinational has pictures of women thoughtfully choosing the appliances in the American mailings. But in the Netherlands, where the same brochure is used (with the text translated into Dutch), there are photographs of a man standing pensively before his appliances.

Everyone is analyzed. African-Americans are especially likely to receive letters offering encyclopedias, as from a history of efforts at self-improvement they buy more encyclopedias and educational reference books than virtually anyone else. Hispanics are likely to receive mailings offering patio furniture—even if they don't have a big yard—because the mailers know Hispanics feel obligated to spend above their income levels on such family-grouping items, however little they might actually want visits from all those relatives. And Chinese-Americans are unlikely to get insurance mailings mentioning old age and death, but instead they will more often receive solicitations emphasizing how insurance could help the next generation.

Often of course the mailings are misdirected or ignored, which produces an awe-inspiring amount of waste. An estimated half trillion items have been sent out during this century in America alone, with perhaps 500 pieces on average to each family in recent years. A suburb of 10,000 people would attract enough junk mail to build Noah's ark each month if all the paper were converted back into the wood from which it came; an American city of 1 million would attract enough to *fill* one ark every three days.

Rich people get more junk mail than anyone else, though here regional and ethnic differences break down. The effect is well known. ("You ever been black?" Larry Holmes, the champion boxer, asked. "I was once, when I was poor.") One sign you're at the top is getting packages with pictures of cars that look best with a château beside them, and what's on sale is the château, not the car. Another, and it's good to be prepared for these things, is getting unso-

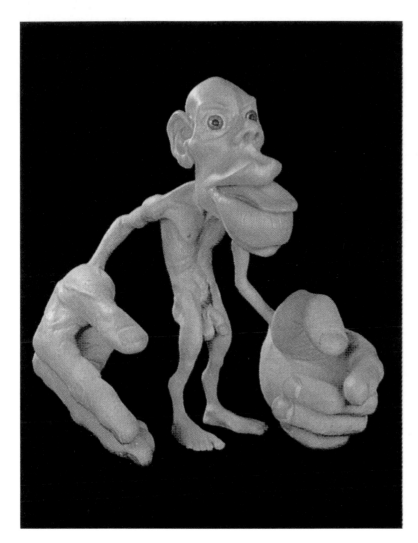

The lumbering shape of a sensory homunculus—a model that indicates how much space our brains give to the nerve signals cabling in from different parts of our bodies. Lips and fingers are loaded with nerve receptors, and so loom large; stomach and legs are more sparsely connected.

licited furniture promotions through your door that show pieces in forest green or burgundy. For some reason, these are the colors which the richest 3 percent of Americans are judged to select most. Do not, under any circumstances, feel complimented if you start getting furniture offerings in sky blue or grass green. The companies must know something from your financial records that you don't, for those are the colors which only the poorest 10 percent of Americans preferentially buy. And to be honest, if you're getting furniture promotions at all, you're not quite there. One British cabinet minister remarked with disgust upon a particular opponent of his, a mere millionaire, who was so deprived that he actually bought furniture. The slur would be lost on many Americans: a true aristocrat is someone who inherits all he needs.

The mailings can even tell you whether you're going up or down. "First-time buyers of investment devices" are likely to have been tracked by their types of cars or their neighborhood history; you and the other readers of that brochure are probably all going up. A mailing announcing "Consolidate those irksome bills through cash!" is a warning sign that the grass green furniture brochure is coming next.

The microwave buzzes, and the Danish—having, with a typical 2.45 MHz motor, been pummeled precisely 147 billion times in its one minute inside—is ready.

It looks delicious, icing glazed and caramelized, steaming with butter-rich vapors. Some people would deduce from this, that it's made of things like fresh icing and flour and butter. But that's no more likely than the family's fresh orange juice being made with fresh oranges. What we call Danish pastries—the Danes call them Viennese bread—couldn't be produced in mass quantities without some substitutions along the way.

The first thing to work on is that icing. It's hard to keep icing sugar enticingly white so the bakery company will just slap some white paint on. That's why dollops of titanium dioxide—the same chemical in the buckets of leftover white latex paint in the garage—form a good part of that gloopily delicious white substance on top. Where some brown, caramel-suggestive swirls are needed, brown waxes, including the indelible rosin used on violin bows, are often used.

Most of the rest of what's inside is pretty well known: flour, sugar, nuts, and oils. An ordinary bakery pecan roll can easily contain more fat than you'd get from a plate of eggs, bacon, and pancakes doused in margarine. But Danish pastry also needs to have what food psychologists (these are not researchers who interview foods—such people exist, but are generally only seen during visiting hours—but rather ones who ask others how they feel about foods) call "enhanced mouth-feel."

The simplest additive here, used in the most inexpensive Danish pastries, is made with dried extract of red algae, bleached so that no red color shows. It's cheap enough, just go to the right beach and there it is, and it does add a little of the desired stickiness. But this algae's quality isn't very high, and can also produce what's politely called "abdominal distention" if eaten in too great a quantity. Processed chicken feathers or the scraped belly stubble from scalded pig carcasses are often added to these lowest-price pastries, as their extracted proteins help in softening the flour that's used.

If you can afford a slightly better grade of Danish pastry there's likely to be a superior algal seaweed inside: the *chondrus* species. It's so much better and so useful for adding smoothness, that some is often used in cosmetics, which is something to think about when chewing a Danish while trying to put on mascara. But it's still nothing compared to the third and highest grade of mouth-feel enhancer. This is the high molecular weight polysaccharide called galactomannan. (Do not read any further if you like expensive Danish pastries.) Galactomannan is produced by processing the pods of a tree which originally grew in the Mediterranean region. It is an immensely sticky substance, ideal for mouth-feel needs. The tree, however, was also cultivated thousands of years ago. And who would have needed something so sticky in those years before it was included in Danish pastries?

When the Pharaohs of ancient Egypt died and had their brains pulled out through their noses and were embalmed, it was important that they be securely wrapped for their voyage into the land of death. This meant thick burial shrouds. What held those burial shrouds together, surviving so well, with a stickiness so undiminished by the centuries that you can still pull out the active molecules from just above the mummified bodies in museums today? It was galactomannan.

Most of the family can't taste any of these extras, and they eat contentedly: nostril skins twisting wide and rib cages convulsively expanding—our heart rate almost always briefly speeds up when we sniff food—to help whirl in the spurts of air needed for proper appreciation of these delicious vapors. About 20 percent of the population however are termed supertasters. Supertasters have more taste buds than other adults—about 9,000 is the average adult figure—and they even have more taste buds than the elevated numbers that children have. Very occasionally a man possesses this anomalous inheritance, but more often the sole supertaster in a family is going to be the wife.

(No one knows why women are better tasters. One idea is that it could be a way of detecting very low-level bacterial infections in food which wouldn't matter normally, but could be important during pregnancies.)

It's a lonely life. You end up spending certain mealtimes asking your husband, in whispers so as not to upset the children, if he's sure, really sure, he hasn't detected that smell you're sure is there. You will smell milk going sour a day or so before anyone else, and notice when someone left onions on the bread board a week ago. No one, ever, will believe you. This sensitivity is even stronger near a woman's monthly period, for the nose's inner lining thins slightly then, and incoming vapors are more acutely detected.

If the parents are distressed at their kids' eagerness in gorging on the Danish, they can take heart from an Ohio study done on children in the hospital. When the children were given their choice of any food to eat, they began as you'd expect, loading up on chocolate cake and other sweets for the first few days. But within a week—so long as there were no reprimands from the nurses—almost all of the children ended up willfully gobbling plain bread and fruit and—even the boys did this!—lots of chewy vegetables.

With the Danish eaten, the final harmony takes shape. The boys are immersed in their aliens-blasting computer game, and the girl has stopped bitterly flicking through her magazine—it's no good, really, if people have stopped paying attention to how dissatisfied you are—and is on the phone to a friend. She's most likely sitting down, for women almost always sit when they phone, whereas men stand, and she's probably holding the phone in her left hand, a curious family-linked trait which left- or right-handed people in a family are equally likely to follow. (One of the only sidedness traits more rigid in a family is which thumb you put on top when you sit with thumbs crossed over folded hands. It's passed on culturally, not genetically, but is usually 100 percent consistent among the kids within one family.)

She might be getting irritated by the distracting noise of her brother and his friend talking, but here there's an easy maneuver to help. If you switch the phone receiver from the left side to the right, you're likely to hear the words you want better. The reason is that the cabling leading in from the right ear goes preferentially to the parts of the brain on the left dealing with language processing. In a noisy setting, when someone is given two words at the same time, broadcast over headphones to his different ears, he usually registers the one coming in from the right best of all. This has some truly odd consequences. Tapping your forefinger is also controlled by cabling that stretches from the opposite side of the brain. This means that if you tap your right forefinger, then some of the language areas on the left are being used up. If you're asked to repeat back what someone says to you, then you won't do as well as if you were tapping your left forefinger instead. (Try it.)

The mother is finally able to return to the newspaper. Silent reading has not always been considered an acceptable way of passing time. A contemporary of St. Ambrose described the spectacle of first seeing someone reading to himself: "When Ambrose was reading, his eyes ran over the page . . . but his voice and tongue were silent. . . . We wondered if he read silently perhaps to protect himself." For almost all of history reading had been something done aloud. Early texts had none of the periods and commas and capital letters and paragraph divisions we're used to for helping out quick, silent scans. Only in the latter seventeenth century, with the spread of private Puritan contemplation, among other things, did Ambrose's strange habit become popular. We still have some holdover of the previous time in the few symbols some-

Eyeballs stretching forward from the brain, revealed in this horizontal scan through the head.

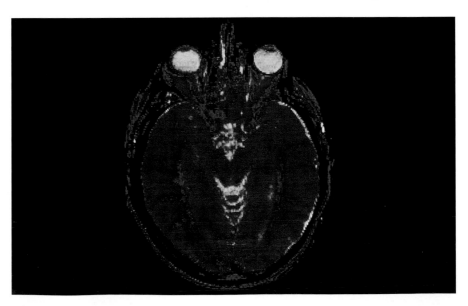

times put into our text to symbolize emotion and to help readers speak the words: the ancient "?" and the "!" Before the mandatory education laws of the 1800s hardly anyone was going to be doing any reading at all. In Spain in 1885 for example, only 500,000 people out of the population of 16 million could read and write. In Egypt and Saudi Arabia even today perhaps only 30 percent of adult women are educated enough to share the blissful morning relaxation of reading a newspaper.

The mother's eye pupils aren't moving smoothly over the words. Rather they hold still over an interesting chunk of text for about 200 milliseconds, then several of the miniature muscle cables digging into her eyeball give first a preliminary tensing and then heave. The eyeball starts spinning at impressive velocity to the right: so fast and so hard, that we find it impossible to see anything at all in these blurry rushing moments. After about 20 milliseconds, decelerating tugs from the other eyeball muscles start, and in another 10 milliseconds the blurry flight is over: the wife will start to process her visual signals again, with the eye neatly aligned with another chunk of text, further along the line. The only certainty is that when she comes to a period all heaves and tugs will cease: we almost always squeeze our eyelids and blink tight then—as again can be demonstrated by trying. Otherwise her blinking holds at a steady 18 or so each minute, though the more difficult an article, the fewer the blinks. All the while, oxygen molecules billions of years old land from the room air, to nourish the cornea sheet resting precariously just below her eye surface.

The dad glances over, and now yet another feature of eyeball movement comes into play. People don't just swell the pupils of their eyes when they look at babies. It's a far more universal sign of pleasure or distaste. When the newspaper page reveals a picture of a politician they like, their pupils briefly spurt wider; the facing photo, of one they dislike, will reverse the lens-rim muscles, and cause the eyelids to flinch. Because it's all unconscious, at times it can be embarrassingly revealing. If there's a young female movie star pictured on the page the husband's eyes are likely to quietly dilate; the wife's eyes will usually do the same for a male star she finds attractive. But men, the silly reflex machines, will usually dilate their pupils on just seeing the *word* "nude" in an accompanying headline, an excess women generally avoid. The sex differences go deeper. Men usually remember having seen a sexy ad but can rarely remember its content. (This is like Churchill's remark that he couldn't say he remembered Latin, but he definitely remembered having studied it.) Women, more sensible creatures, turn out to remember both ad and content, however much a hunk within it might have made them demurely dilate.

For both sexes almost everything that's read is going to be forgotten. An ordinary morning's newspaper can have 15,000 words or more, and if you add on all the office memos, backs of cereal packages, road signs, TV listings, mall ads, and even, for when we're really bored, the tiny print in the magazine ads: from all that our eyes might be faced with 100,000 words in a week. The daily *New York Times* alone, estimates have it, has more bits of information than the average seventeenth-century person would have come across in a lifetime. It adds up to 5 million fresh words each year; 50 million words—vast torrents of potential knowledge—each decade.

It's too much. The incoming letters get processed in the visual centers of the brain, broken into their component parts of angles and curves, sent swirling in quick, high-speed circuits to the advanced reasoning centers of our cortex, and then unceremoniously dumped. Only a very few signals, the ones that get further shunted through the region of the brain called the hippocampus, have a chance of becoming embedded in long-term storage. (After a stroke in this region people can end up living in a permanent present, utterly surprised at the presence of a visitor with whom they might have shaken hands and chatted just a minute before.)

The boys yell as an alien battle cruiser explodes, and the dad switches attention to the computer game they've been playing. The ten-year-old's brain is near its peak, pumping away with about 50 percent more oxygen fuel than the brains of the parents. The dad awkwardly tries to work the controls, but this isn't easy. University of California researchers have taken brain scans of people playing computer games. Beginners of whatever age go through desperate surges of neural glucose metabolism as they try to recognize what's going on. Experienced players don't, and can skillfully glide along with minimal brain exhaustion. Children even—this is what's especially unfair—are likely to have started growing different brains from their parents. Violinists end up with permanently enlarged clusters of cells in the cortical regions that control their left hands. (This is the hand that does the delicate fingering;

the right pretty much just saws away.) The younger they start, the more that region grows. This is a further reason parents suffer, while kids, with their analogous control-center enlargements, have to learn to be polite and wait.

The kids shouldn't be too cocky though. Out-of-date computers are often disparaged and the fact that Apollo 11 went to the moon with an on-board computer carrying only 16K of memory is regularly ridiculed in the computer press. But how many of today's young terrors with galactic blasters could program a spacecraft that would navigate to the moon with just 16K of memory?

Higher up, clouds of nitrogen dioxide left over from the earlier pancake cooking are dissolving harmlessly in the flecks of water within the moisture clouds the family breathes out, floating as invisible cumulo-nimbus formations over the breakfast table. Farther inside, within the nostrils them-selves, miniature rainfalls of sulfuric acid are taking place. This is because everyone's breathed in slight amounts of sulfur dioxide, which floated up

A group of X chromosomes, its fragile genetic information constantly being repaired. Individuals who lack a corner of the X in one chromosome have to get by with a smaller Y-shaped one—these misfits are called males.

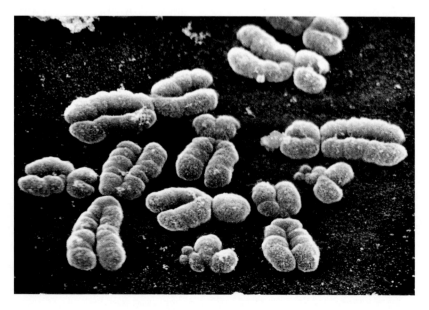

from the raisins in the breakfast cereal, where it was used as a preservative. Now it reacts with more of the moisture coming up from their lungs, to be harmlessly dispersed inside them. The kitchen's walls can themselves protect the family, for all the talking and plant watering and coffee-making lays down fine films of water on these walls. The abrasive nitrogen or sulfur compounds that land on the walls around the family are trapped. They bubble, dissolve, and then—twenty to sixty minutes after contact, depending on quantity—they're gone.

Detergent residues on everyone's clothes are another invisible menace. Skin defends by bringing up extra oil barriers, as well as replacement salts and amino acids as needed, to repair the local damage. Washing machines in America are generally worse than the best washers in Europe in this respect. The ideal washing machine is one that works with low suds—too many foamy suds and the sideways bashing that knocks the dirt out is cushioned—and has sturdy enough pumps to drain them all away. Machines that lack the drainage power, leave the skin-attacking residues. (When a wool cycle has been selected, an embarrassingly simple alternate mechanism operates. On most machines the wool cycle merely means that less water is poured in. The clothes are less soggy, weigh less, and so bang into each other with less thread-damaging whomps.)

Outside, the little patch of sunlight that's reached the patio is doing a good job of cleaning the air the family is going to breathe, traveling in through the open window. Ultraviolet rays kill thousands of live microbes floating in from out there. One reason we have fewer colds in summer months is that there are more daylight hours for ultraviolet rays to sterilize the air this way. The unobstructed sunlight also contains a fleet of tiny hydroxyl molecules that act as hovering garbage collectors, skimming along at window height and below, scooping up the methane leaking out from the family's plastic-lined garbage cans, as well as much of the carbon monoxide that's made it past the front lawn, and drifted around to this side of the house from the street.

Even the farthest traveled of all intruders are being quietly dealt with now. These are the cosmic rays slipping down from the ceiling, many launched from stars immensely farther away than our sun. A few are simply absorbed in the ceiling paint, seemingly exhausted after a flight of many thousands of years. The rest are still arriving fast enough to batter everyone's eye lens cells, which isn't too bad, as well as the crucial DNA molecules farther inside, which is. Without help your kids would soon look seriously strange, and the mutation rate would be too high for us to survive. Luckily, even as the wife's eyes jump to a new viewing angle on her mashed-wood page, repair molecules inside everyone around the table start snipping out the damage. Proofreading chemicals carry the broken fragments away, and while those are being replaced, the damaged bits are either reused, or are pumped toward the bladder, where they will end up further enriching the already aluminum-laden urine stored within. The outer space bombardment is held back a little by the earth's magnetic field, which stretches like an invisible arced dome over the home. Vast fleets of the incoming particles are constantly dragged away, out of the apparently empty sky, before they can get in and led to distant snowfields and ice near the Poles, where this magnetic field reaches the ground. Only in periods when the magnetic shield weakens—it seems to happen every few hundred thousand years—will the proofreading DNA repairs be overwhelmed.

The baby, meanwhile, is crawling along the hallway. He listens contentedly to all the sounds the adults don't hear: the air whistling over the polished front hall, the mail slot clacking, and the dog's breathing upstairs. There's a long moment's pause, then a final glance back to confirm no one is watching.

This journey will have to be fast, for a baby is still too insulated by fat to keep up high-speed movement for long without overheating. Its tiny 1.4-ounce heart begins to contract more powerfully, the eye-protecting blink rate goes up, and for the sake of keeping all its brain cells powered-up at an optimal seventy millivolts, its body sends spurting hormone signals to mobilize food sugars. Only then—with a final quick oxygen intake to make sure—does this portable exploration machine, little blanket gripped in one hand, set out.

It's slippery on the tiled floor, but saltwater squelches from the 27,000 or so sweat glands the baby has on each palm—a density three times higher than that of the adults so unadventurously staying behind. Thin ridges of skin—the fingerprints—sticking out on the bony extensions from those palms become hundreds of suction cups to help in giving a nonskid grip. In most families the direction of the fingerprint loops on the baby's and everyone else's middle finger will be aimed toward the little finger; in about 10 percent of families it will be consistently different and aimed toward the thumb. Finer details vary even within a given family, which is how the police measure of individual fingerprinting began. French police authorities were using it by the 1890s, though Scotland Yard is proud that a British investigator, Edward Henry, used them at least two decades earlier to control pension fraud in the army. For his efforts Mr. Henry received a baronetcy, and was popularly known as Mr. Fingertips. Chinese historians would be less than impressed: Middle Kingdom bureaucrats were using fingerprints over 1,000 years before that, as something like personalized credit cards for legal identification. There are similarly unique ridge patterns on toes, palms, and even on the roof of everyone's mouth, but those are less often left at the scene of a crime.

More cosmic rays are leaking down from the ceiling, but the infant's baby-size DNA repair systems activate. The tile's surface reflects bright light, which electric reception units in its eyes amplify and channel in for processing. The rush of colors and sensations could be confusing, but an exploring baby has been designed to do a lot of presorting. It won't worry about two light signals that are vibrating bare millionths of an inch apart, if they're both within the blue range. Yet two signals that close which happen to straddle the color boundaries its parents are teaching it—one perhaps on the very edge of blue; the other in the adjacent green realm—will be identified so the baby can concentrate on them. It peeks out the mail flap as fragments of Isaac Newton come floating up the stairs. This isn't because it's a house bought from a wild-eyed Stephen King–like real estate agent, giving the parents such a special price and answering with a haunting laugh when they asked why it was on the market at such a bargain. Sir Isaac is actually floating up the stairs of every family's house this morning, as always. The reason is that the human body contains at least 10^{25} nitrogen atoms, and long after a person's life has ended, a sufficient number of those molecules filter into the atmosphere to drift into almost every parcel of air. This baby's distant ancestors are rolling up with Sir Isaac too, finally to meet (and be breathed in by) their progeny. Believers in the transmigration of souls might take note. (There's also a certain amount of your own self always coming back, for every nine years or so almost every single molecule that makes you has gone, either

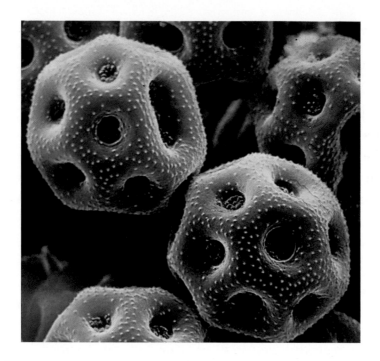

Pollen from a rose, 8,000 times life size, showing the pitted surface that blindly mates with flower or nose.

floated away or poured out. This solid stuff that was you doesn't stay entirely dispersed, and in its random travels some will steadily—in small parts—come rolling back home too.)

The baby looks up and glimpses the bright blue sky. It's blue because almost all the ancient photons which the dad was soaked with are absorbed by distant dust particles, and only a few—the ones moving with the frequency we call blue—rebound to reach our eyes. The baby stares at the sky-dome spectacle, blinking a little faster than usual as some of that otherwise invisible dust swirls in. Fragments $2/1000$ mm wide of distant forest fires, weathering mountains, and dust storms on the other side of the planet all land on his eyes, are squeezed over to the lachrymal duct pink areas and sent down for disposal in the nose and then stomach once again.

Bobbing in the morning air are live mating capsules, which have the habit of landing inside any watching human's nose, however young or small, and once in there, trying to mate. They quickly dissolve their uppermost surfaces, and dribble out enzymes. Then, still in the first few moments, they start to push forward their protein-rich hydraulic tubes.

Some people don't mind at all when pollen does this to them, and will venture out amidst the daisies and the hay fields, ready to enjoy a full, nose-lowered satisfying inhalation of air. Other people, however, would not leave their homes without tight-fitting gas masks on. Unfortunately both are likely to be in the same family, which is why shared decisions about summertime outings can be rough. To some extent there are genes that hay fever–suffering parents pass on, and which load their children with the reflex to make excessive amounts of the IgE antibody that triggers histamine explosions when pollen arrives. But not everyone in a family is equally likely to suffer. Oldest kids get hay fever four times as often as last borns. (It could be that they picked up fewer colds, as the family had fewer members when they were young, and that reduced their abilities to control any overeager IgE, but nobody knows for sure.)

Ozone gusting forward is more of a problem. The family members back inside are protected, for over the years their tracheal linings have become used to the usual levels of ozone floating up to this house. A baby hasn't had time to develop that protection. A lot sticks to his terry-cloth clothes. Ozone attaches a hundred times more readily to cotton than to glass or plastic and so is gushing directly faceward. The incoming ozone sinks into the sticky coatings on its breathing tubes, and there's an ever so faint tickling sensation as some of the cilia scaffolding underneath begins to collapse. In most parts of our body such slight tickling would be no problem, but the nasal mucosa always have to be protected. The sensitive endings of the trigeminal nerve pick up what the baby's receiving and send it through a direct conduit—there are special holes left open in the skull for it to fit through—to the brain. There's a brief moment of stillness, as quick cerebral evaluations take place; then the baby's face contorts in puzzlement, as the blast of sneezing air starts flying up. The vocal cords narrow, to force the air faster through the throat exit tunnel, and then, often with spectacular nasal-erupting force, out it all explodes.

Oooph! The baby plops down from the sneeze's recoil. Its small distance from the floor helps it land at a safely low acceleration, as does its relatively small internal volume. It's a general principle in the animal kingdom that smaller creatures are safer in a fall. As Gip Wells put it, a mouse can be dropped down a dry well and walk away, a human would get a broken leg, a horse would splatter. And there are advantages now, for from its sitting position, and nicely energized by the fresh air, the baby can contentedly explore some more, looking to see what's interesting here. The previous day's newspapers and junk mail left piled by the door are only worth a brief taste; far more interesting is that inviting electric socket beside the table where the phone books are kept. The baby lets go of its little blanket, then runs its thumb on the cool plastic surface, trying to push into the indentation where the powerful electric charges are connected, live, to the distant power generation station. Tiny droplets of water and gas are hovering in the air near the socket, attracted by the leaking electrical field. Because those droplets have radon-decay products stuck on them, the baby breathes in a quantity of radioactive products almost as great as what it took in under the sink. All electrical cables and gadgets seem to attract these radon-loaded invisible clouds: dishwashers, clothes washers, and bedside clocks are especially surrounded by them. Luckily the levels are high only if you lean right next to the cables or machinery, so the baby's natural defenses are sufficient. (It's possible that overhead power cables accumulate even more of these radon-loaded clouds, but the evidence is still being evaluated.)

Nothing much happens at the socket, so the baby turns to the interesting odor of his sister's leather-shiny black coat. Some of the smell is from the fragments of tobacco dropped from the cigarettes she secretly keeps zipped away in an inside pocket; more of it though is a long-lasting breakdown product of the nicotine she heats up when she smokes.

That nicotine doesn't last long—only a few minutes in the open air—but the cotinine chemical part of it is much more stable. Thousands of the cotinine molecules are tumbling down the sleeve of her coat even now, hours after she smoked with friends last night. They bounce in easy slow motion over her empty sleeve to float to the sniffing baby. Some will go into his bloodstream and end up, months from now, stored inside his growing hair. Even if the sister doesn't smoke much, but just hangs out around kids who do, some of this cotinine will end up there: the children of nonsmokers who are exposed to passive smoking almost always register some of it in their hair.

The baby reaches into the jacket, its fingers slipping along the shiny nylon inner lining. It's cool to touch because the vibrating molecules on his fingertips don't transmit their energy very well into nylon. That's also what makes nylon a good electrical insulator, and why it's likely to have been in the plastic wall socket. There's nylon all over the family's surroundings: door hinges, shoe heels, couch coverings, eyeglass frames, garden hoses, Velcro fasteners, combs, and car parts all are likely to have some of that same stuff in them. One might think that the man who invented such a ubiquitous substance would have been justly proud, but the scientist responsible, Wallace Hume Carothers, nylon's creator, was a tormented man who never believed he had done enough. In 1930 he created artificial rubber, but when this seemed to have no application, he started to work even harder.

In 1935 he came up with the polymer blend based on chains of six carbon atoms that became nylon. A number of name changes though, along with delays in getting a patent, increased Carothers's feelings that he was a failure. It didn't matter that nylon's ultimate commercial launch, on May 15, 1940, was the greatest success in Du Pont's history, with over 4 million pairs of nylon stockings sold in the first week. Nor did it matter that his artificial rubber discovery became crucial to American success in World War II, after Japan blocked rubber imports; nor even that, as one recent review noted, Carothers had created the fundamental understanding of polymers which now occupies the research of perhaps half the chemists in the world. Two years before nylon's commercial launch—a bare three weeks before its patent was finally filed—Carothers filled a glass with lemon juice, added potassium cyanide, and committed suicide by drinking it.

From a pocket of the girl's coat falls a small wood fiber box. This is interesting! The baby lifts the box, and little sticks with bright red tips tumble to the floor. It tries eating one, but the taste is not very good; it lifts up another one now and ponders.

At this point one investigating parent comes out, and sees the baby dangerously near the matches. The baby's picked up, away from this danger, and there are calls, angry ones, to the kitchen. The girl emerges, whispers hurriedly to her phone friend that she's got to go, then races up the stairs, sobbing. The boy and his friend glance up vaguely from their computer game, surprised at the suddenly empty table. They don't quite know why, but one thing is clear.

This breakfast is now over.

3

around the house

An hour later and the baby's happy, snug in his mother's lap, at her desk upstairs with the computer, sipping some nice warm milk from his blue plastic cup and watching her work.

Why is this so nice? There's the comfort of being held, and the murmuring of ancient counting rhymes—"eenie-meenie meinie-moe," meaning "one-two three-four," may have been carried over into English from the languages of pre-Roman Britain. But there are also the amino acids in the milk, including the one known as tryptophan. Normally only a little tryptophan will work its way into the brain, for there's no reason for it to be selected for cranial entry more than any of the other amino acids in milk. But if sugar or sweet chocolate has been mixed in, then the baby's pancreas squirts out extra insulin as it starts to drink. That helps hold back the majority of amino acids, and the tryptophan becomes dominant and has a free way up, up, through the pump-selection capillaries leading into the brain. Minutes after this skillfully dosed sugar and milk mix is in, it transforms into puddles of the powerful

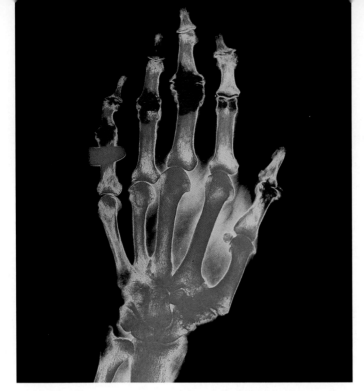

A woman's hand
with the joints
swollen by
arthritis.

neurotransmitter serotonin. That acts like a miniature Prozac dose, leading to a smooth, tranquillity-inducing feeling. It works for adults as well as babies, which is why the old folk remedy of some honey in warm milk before bedtime makes sense.

A vast human hand reaches over the mellowed-out baby, stretching for its own midmorning sustenance; the wife's ring glinting in the computer-screen light. The gold of a wedding band is quite as old as the blizzard of immensely ancient oxygen molecules her hand is passing through. It was created in the supernova explosions of ancient stars, far more quickly than the iron atoms which had more slowly built up inside those stars, and fell toward earth as a fine glinting haze over 4.5 billion years ago. Any diamond on an adjacent engagement ring or, what's more likely till the mortgage is paid for, any preposterously overpriced tiny sliver of a diamond, is nowhere near as old as the star-flung gold or oxygen. Diamonds are crude newcomers, resident on our planet for a bare billion years or so; there are even some later arrivals, gate-crashers really, not even 900 million years old actually available in the shops. Many were created fairly near each other, and only moved upward in rough geological synchrony, in the subterranean super-plume episode of the mid-Cretaceous, when immense volumes of our planet's lower layers were slowly lifted skyward.

The finger which bears these much-traveled rocks is relatively shorter in women than men. Women are likely to have ring fingers about the same length as their index fingers; for men, or at least 70 percent of them, the ring fingers are longer, as an examination of the baby—the relative sizes hold from birth—is likely to show. The reason the fourth finger is selected for these adornments could be a holdover from ancient Egyptian beliefs that this finger, alone, had a favorite circuit through the body up to the heart.

The marriage which led to this consistently placed ring is not always seen the same way in different cultures. Polls show that Americans are the least cynical of nations here: when lists of possible reasons to be married are given—to have kids, for financial security—they insist it's none of those, but say marriage is good as an end in itself. It's the Europeans who are less impressed: in France and Germany today, large majorities say that married people end up no happier than unmarried ones. Britain seems to waver between the extremes. Charles Darwin, famously, made a less than romantic list of pros and cons while he was courting: marriage would mean having "to visit relatives . . . anxiety and responsibility," he wrote; but it would also, he noted in a parallel column, mean he'd gain "someone to play with—better than a dog anyhow." (Boldness won out, and with just seven months more procrastination, Charles popped the question, beginning what actually turned out to be a happy, decades-long marriage.)

It's no help this morning that few people can read a computer screen as well as they can read from paper. Our pupils constantly leap off course, distracted by random flickers on the screen, with the result that we go about 25 to 30 percent more slowly than on paper. A further problem is that reading from a computer screen exhaustingly dries out the eyes. It's common practice to sit straight ahead at a computer, which is odd, because no one reads a newspaper or book like that. We bend. Here before the screen, staring ahead makes the eyes bulge. More moist surface gets exposed to open air, which sucks up fluid from the eyeballs, and disperses it uselessly into the floating room air.

The wife would love it if she had more free time, though according to specialist instructors in time management she already does, but merely doesn't *recognize* that she does. One New York–based course inspired a female executive to gain extra time by no longer waiting, uselessly, passively, during that full minute each day she was stuck in the shower waiting for her hair conditioner to soak in. Instead she started using that time for flossing her teeth, to her immense satisfaction: "I've never had such healthy gums and glossy hair," she reported in delight. Another executive realized valuable moments during the flip-turn at the end of a swimming pool were being wasted. The solution here was to tape up waterproofed sheets of poetry, so poems could be memorized while swimming laps. Being pressed for time is not a development unique to the late-twentieth century. Only for the few decades before the mid-1970s were men's incomes generally high enough so that their wives could stay at home. Before that, women almost always had to work outside the home and fit everything else—kids, spouse, running the house, and maybe even fragments of time for themselves—into the brief intervals left over.

Like the baby, the mother is also drinking a cup of warm liquid. It's a concoction that she would be appalled to learn resembles mashed insect brains, chemically speaking. Early in the Carboniferous period, cockroaches and other primitive insects were already using two types of transmission chemicals in their brains: one, similar to our adrenaline compounds, would speed up the communication of signals in their bulge-eyed heads, and the other, similar to the baby's mellowing-out serotonin, would slow the signals down. For a long time no plant could artificially duplicate these neurotransmitters, but with the establishment of flowering shrubs about 70 million years ago that began to change. Many of the new species had cell mechanisms complex enough to build exact copies of the complex ring structure which insects and other grazing predators used. The result was the plant substances called the alkaloids, which include strychnine, morphine, and quinine. All are modified versions of insect nerve transmitters.

One of these alkaloids worked by being so similar to the original chemical which slowed down the insect's thought transmissions, that it would lock into that chemical's place in the brain of any creature that consumed it. The creature's own slowdown signals would no longer work, and the brain of any predatory insect that still insisted on seeking out the plant would go haywire. The plants which produced this alkaloid concentrated it, and dangled it down to be deposited just where the predators might come, the better to ward them off. It's the dangled concentrates—better known as coffee beans, of course— that we consume in the morning.

Coffee first became known in Europe during the seventeenth-century Turkish siege of Vienna, and spread so quickly that within a decade Pietro della Valle in Italy was complaining at how excessively "students who wish to read into the late hours are fond of it." By 1650 the first coffee houses reached England, and the oldest one is still there, located at Queen's Lane, Oxford (which is where these words are being typed, encouraged by the ancient plant extracts). Americans aren't as prolific consumers of coffee as the Scandinavians, who are world champions with an average yearly dose of 612 cups each, but they still manage to pour down about 33 million gallons of it each day, or the equivalent of thirty seconds' worth of what flows over Niagara Falls.

As the wife is occupied in perfecting her brain chemicals, the baby has sidled down to the floor, and is examining the curious little paper squares it has lifted from her desktop. It doesn't recognize them, and so, as it does with virtually all unknown objects at this age, pushes a few into the surefire chemical analysis unit of its mouth to get further information. There's definitely some sort of chemical coating on one side of the paper, quickly melting away and leaving a fun liquid feeling on its tongue. But with a further tongue-rubbed sampling there's now something reminiscent of this morning's baby food extruding from it, and at that the baby quickly makes a face and takes the square out; holding the wettened mass for a safer inspection in its hand.

Postage stamps are an intricate layered sandwich of chemicals. The glue is a true masterpiece of the chemist's art. Think of the problem. Not only does the glue have to be sticky enough to hold onto an envelope, but it has to be not *so* sticky that it grabs permanently on to your tongue. It has to do this with only the amount of saliva we're happy to dribble off, and then it has to stick to the envelope firmly, but still give you a moment or two to readjust its position. Finally, even once the chemists have worked out something that's tongue-attractive and humidity-resistant and briefly free-sliding, they still have to throw it out if it tastes bad. Or offends anybody's religion. Or is too expensive. Or too high in calories.

The solution is to mix baby food with Elmer's glue. The Elmer's—or similar petroleum-derived polyvinyls—gives it a sturdy hold, while the powdered plant starches of the baby food gives the concerned user time to correct any sideways-skidded placements. It's true that a little of the mix sticks to your tongue and unglues itself a few seconds or minutes later, so that you always end up swallowing a little bit. But that's no problem. The starch is easily digested, coming as it does from healthful corn in America or potatoes in Europe. The Elmer's admittedly is less nutritious, as it balls up into clumps of white glue in your esophagus and stomach, but at least those are microscopically tiny and in time dissolve. Dieters can be assured that even the most thorough licking of the back of a typical U.S. postage stamp will only give them 5.9 calories. Larger stamps can carry greater amounts, up to a gut-busting 14.8 calories, but convenient sponge-moisteners are usually available to avoid any between-meals temptations.

All stamp lickers, and not just thoroughly testing babies, pick up a further sequence of odd chemicals when they lick a stamp. These arise because stamps are normally stored in big sheets, and what's on the top of one migrates, at least in part, to the eventually to-be-licked gummy bottom of the stamp sheet above. There's usually a tiny amount of the embalming fluid formaldehyde, just as in the vapors from the under-sink cupboard downstairs and in the orange juice. In the kitchen it stiffened the bonds between orange pulp, or between wood or plastic grains; in the stamp it's put in to strengthen the soft, thin paper used. What seeps out is actually excellent at killing any bacteria that might be tempted to try to colonize the abundant starch fields of the glue; quality-control specialists will only have to add other bactericides for this if they find not enough formaldehyde is getting through. A little of the clay and chalk from the paper will have slid out onto the glue, too. Then there's some of the optical brighteners used in laundry detergents—on the stamp paper it adds brightness, too—and even some of the algae extracts

A guitar string, magnified 55 times. The core nylon strands are wrapped in a thin steel tubing.

used for moistness, as in the day's Danish pastry. All this floats up with each eager tongue swipe, to get swallowed with the corn granules and polyvinyl glue.

Back at the desk, the wife dawdles some more, taking another careful sip of coffee, then deciding she really should have a section of the grapefruit that's on her tray beside the coffee cup. This is good. The coffee she's been drinking may contain the strange chemical called kahweol, in addition to the caffeine mash, and kahweol, unfortunately, makes cholesterol levels go up in anyone who swallows it. Grapefruit will counter that, for grapefruit is rich in pectin, which binds away blood cholesterol, especially if one's bold enough to really dig into the pulpy membranes where the pectin is densest, ignoring, for the true pith scrapers, those troublesome spatterings on distant window-panes. Individuals who can't bear grapefruit, and who fear those awful pithy bits, can get around the problem by drinking only filtered coffee. The kahweol molecules stick to filter paper and are left harmlessly behind. Both are solutions that any cholesterol-anxious dad downstairs could have considered.

Her delays are tormenting the primitive silicon brain that waits inside her computer, and has been programmed to always, however much it's re-buffed, try to keep in contact with the World Outside. If it has an alarm func-tion set for 7:00 A.M. next Monday, then at every second this weekend its program will consult the onboard clock to see what time it is, and if it's not yet 7:00 A.M. on Monday, which it won't be, not for another 20,000 seconds or more, it will accept this disappointment, calmly waiting, till its computerized instructions build up, and then, ever hopeful, it goes ahead, having totally forgotten what it's just learned, and checks again.

Unaware of these waiting shepherds, the wife lifts an oblong plastic box to her ear.

She'll waste a little more time by phoning a friend.

Most people think the ringing sound they hear after dialing is produced by the phone they've contacted, but that isn't so. The small microphone in its handset isn't working yet, and what you hear is simply a pretend ring, sent back to you from the phone company's switching center. But then the friend answers, and all worries are forgotten: heart rates and breathing almost al-ways briefly speed up at the excited pleasure of making this contact. There's little question it'll be a woman who answers it: when a man and a woman are the same distance from a ringing phone, it's far more likely that the woman will go answer it. (The most detailed studies have been in Britain, where it's three times as likely.)

The skill of dialing that starts off this whole operation is now wide-spread, but once it wasn't. Dwight Eisenhower lifted up a phone one day after he retired from the presidency, in early 1961, and had no idea what to do when an operator didn't answer. He had last used an ordinary phone decades before, when dialing was still rare: in all the time since, official aides had seen that his calls went through. A holdover from that era of rotary-dial phones is the system of area codes Americans still use. The biggest cities have area codes with the lowest digits because low numbers were the quickest to dial. Thus Manhattan's 212, Chicago's 312, and L.A.'s 213.

The baby looks up from its position on the floor where it has been contentedly occupied chewing the bottom of the bedspread and only occasionally ducking its head in self-defense as grapefruit splashed, wondering why his mother is delaying again. When she sees him look to her she smiles, and he races away, crawling in fun toward the parent's clothes closet, door invitingly open, from which emanate the odors of cedar walls and fine cotton shirts and, especially today—after work on Friday being when everything is collected from the cleaners—the tangy, rich vapor cloud from all the freshly dry-cleaned clothes.

It's hard for a baby or its watching mother, or anyone without a large gas chromatography unit strapped on to their eyes, to see the vast number of toxic molecules in that cloud, invisibly billowing toward them. This is consoling, for it's powerful stuff. Dry cleaning isn't actually dry. Almost everything you leave at a dry cleaners has to be immersed in a solvent to get the stains off. Afterward the sopping clothes need to be dried, but even though the staff at dry cleaning stores do manage to get almost all the moisture off, a certain number of chemicals remain. The most important of these is the potent substance called percloroethylene, commonly known as "perc."

Bring your dry-cleaned clothes home now, and the perc is already vigorously outgassing from them. Its vapors create a high-pressure cloud inside your clothes closet, stabilizing when it's reached about 18,000 micrograms of the molecules in each cubic yard of the air there. Keeping the door closed helps trap the stuff inside, even though a little will seep out from underneath the door. But who can keep their closet door closed forever? Who, indeed, can survive on those flutters of quick openings and then quite unconcerned just-as-quick closings which preposterously lucky beings who suffer no dress-code tyranny—that is, male human beings, who can put on the same suit, day after day, for *years,* and still not be considered dowdy—find sufficient? When it's opened wide, the molecules aren't trapped at all. Combined with the constant underdoor hissing escape, the perc molecules reach levels that average 1,200 micrograms per cubic yard in the bedroom's air and 450 micrograms in rooms farther away in the house.

Once it's inhaled our body's enzyme systems turn the cloying perc into other chemicals. But this only makes it worse, for those breakdown products include such dangerous items as vinyl chloride, chloroacetic acid, and, especially phosgene—the old World War I poison gas. Those get dumped in the baby's bloodstream now, to join the smaller amounts it received, along with the mother, when they had been sitting farther away. Some of the breakdown products are breathed out almost as fast you breathe the raw perc in, but a number of them accumulate in the body, especially in fatty tissue or in breast milk. Anything up to 50 or even 100 ppm in our blood isn't too bad, as we can urinate that away a few hours after breathing it in. It's the amount above that—and how long have you slept with it seeping from an open closet door?—which build up and get lodged inside our body to stay for days. You can measure the buildup in dry-cleaning workers: they don't have much on Monday morning, but by the afternoon the levels are rising and on Friday they are at a peak, taking all weekend to clear away. (The result can be more miscarriages and deformed sperm among dry-cleaning workers, at least in Scandinavia where this has been studied.) Even going shopping in a grocery near a dry cleaner's can bring you some, for the perc sinks into fatty foods such as butter or margarine—studies have found it drifting into packages of food, just from dry cleaning being carried home in a car.

Leaving your dry-cleaned clothes outside to air might seem helpful, but isn't very effective. After six hours of airing the perc level only goes down 20 percent. Emigration is a further possibility, for in Germany and several American states perc is being phased out. Until then though, you can check with the regulators to see that your dry cleaner is one who finishes the drying of your clothes properly: there can be a 50 percent or greater difference in how many of the perc molecules are left on a fabric.

The phone call's done, and the mother lifts her baby from the closet entrance now, her breath blowing extra perc molecules toward his tender little nose and eyes as she bends. She settles back at her chair, the baby on her lap again, and rummages through the cables to set up her laser printer. The baby watches fascinated, and makes fumbling movements, eager to join in. But will she let him help explore?

Different cultures handle this in different ways. In one survey, American parents said what they most want for their kids is for them to be independent and self-reliant. This is fair enough, as it matched what American adults wanted for themselves. In Indonesia and the Philippines however, barely one-fifth of parents went along with that; over 80 percent said quiet obedience was the most important trait to instill in kids. There's a similar difference in Japan, as studies by patient graduate students observing Japanese and American parents at home have shown. When a baby is edging forward eagerly on the parent's lap, the Japanese mothers try to settle it back with soothing or lulling sounds. The American mothers are more likely to start actively chatting with the baby, and encouraging it to explore. Girl babies get it worst of all. The Japanese mothers don't chat to them as much as they do to male babies; the watching students were startled to find that mothers didn't feed or touch female babies or even change their diapers as much. (Lest the American reader feel entirely superior, an experiment was conducted which kept a close look on maternity wards. When both parents were in the room, American mothers were just as attentive to their newborn boys as to the girls. But let the father leave, and then—despite later protestations of total equality!—the girl babies invariably got more cooing, hugging, and general contact.)

In this exploration-encouraging family, the baby's hand is led forward to switch the printer on, then the desktop copier is turned on, too. Tiny clouds of selenium and cadmium dioxide spray over the baby as the motors heat up; then some nitrogen oxide and a little carbon monoxide from the copier's toner; then finally, as the high-voltage electric discharges inside the main motors really get going, a full exhaust blast of corrosive ozone molecules overpowers them all. This is why people who have to use an office copier a lot in an enclosed room end up feeling bad. The baby has had enough, and however much its mother can't understand—for with her adaptations the ozone is no problem, at least now at at these low doses—wriggles desperately to get down. It pads away quickly on little palms and pajamaed knees, squeezes through the half-open door to the upstairs hall, and is gone.

Downstairs meanwhile, a floor below all these activities, the dad is coming in from outside; his lawn mowing finished, or nearly enough. He knows that certain neighbors will have problems with the jungle-like patches left at the far end, but their concerns are easily put aside, for there's a sports program on television, and priorities have to be understood. He strides toward the living room, holding a freshly cut rose in his hand. Desperate chemical messages are still screaming out from the dying plant, leaking behind him in the air, but there are no circling wasp predators to call here; the revenge calls just stream free in utter uselessness, adding though, a little, to the general fresh plant smell.

The living room door is left open as he hunts for a vase. There's some dust on the shelf where the vase is located, and part of him, the sensitive house-caring part, knows he should go ahead and clean it, but another part of him, the male part, realizes that's nuts.

He stands back to get a proper look. It's always been a puzzle how even high surfaces get so dusty, but even at this moment he's adding to it. Unless you live near a dirty power station or big road, most of your household dust will simply be chunks of your family's own skin falling off. Each of us is layered with about five pounds of the stuff—more for the larger members, less of course for the baby—and within a few weeks almost all of the top layers are pushed up and off, as fresh layers sprout up from below. (Most of the growth upward takes place at night, as we sleep.) From just five people growing the stuff and steadily popping it loose, there will be many pounds of skin launched each year. Divided into several billion separate cells, the dried skin fragments flutter everywhere the air currents blow them in the house, arriving downstairs in just half a day of ordinary floating. Any ledges that stick out from a wall are especially good at collecting a terrific biological history, because the stately in-house air seas carrying our skin don't just circulate at random. In this living room, as in most rooms, air rises up in the middle of the room, then spatters sideways along the warm ceiling till it reaches the cooler walls. There, finally, it gets led slowly down, bringing its tiny skin cargo, while anything sticking out—shelves will do, but window ledges are also excellent—is buried under the full impact.

The father sits on the couch in front of the TV. He squeezes an infrared-squirting object—the remote control—and the TV immediately comes on, which is no surprise to the electricians who made it, as a modern TV's circuitry doesn't need the long warming up of earlier models. When the remote control's signal arrives, a switch is released and the electromagnetic signals soon start bombing from the screen into his retinas.

And this is bliss.

A metabolic state unique to television watching now begins: the father's eyes remain busy, constantly scanning, but the rest of his body is slowing, pulse and temperature and muscle tension all decreasing in the average viewer to a level that steadies at 13 percent down from even just ordinary sitting. Any truly intent TV watcher will soon be in the extraordinary position—as physiological measures will confirm—of burning fewer calories than if he were doing nothing at all! This TV body-crash is disheartening for anyone who expected that his body's usual rate of metabolism would use up many of the calories from his last meal, but it's worst for anyone who's overweight to begin with. A check of thirty-one girls watching old reruns of *The Wonder Years*, breathing tubes and other physiological measuring devices attached, found that it was the chubby ones whose slowdown was the greatest.

Are there at least some health benefits from the decrease in heartbeats? A slowing of 10 percent will, after all, save seven heartbeats a minute, which is 420 an hour, or, in three hours of this blissful sports isolation each weekend, a third of a million fewer heartbeats each year. Unfortunately, it's not quite a justification for massive slumping. A man's heart rate is already lower than his wife's, even when he's not lost in TV oblivion. There's a general physiological law that large mammals invariably have lower heart rates than smaller ones. Men's hearts beat more slowly than women's, and women's hearts are slower than kids', whose are slower than dogs, whose are slower than cats, whose are slower than mice. But although this means that the wife pounds out several million more heartbeats each year than her husband, her risk of heart disease is still less than his, because of the general protection from arterial clogging she probably gets from her hormones.

Because the couch-dweller's eyeballs don't join the general slowdown, a condition very similar to a waking dream begins. There's the same inertness of general body muscles; the same exception for the six orbital cavity mus-

Turn on a television or radio and a 740-mph sound wave blasts out, expanding as a perfect bubble.

cles spinning and tugging the distantly watching eyeballs; the same sudden spurts in pupil diameter as something of special interest appears in the visual field to be tracked. TV shows are a perfect match: like dreams they bring us a world where cars start without fueling up, guns fire without reloading, and you can jump from one scene to another, hair unmussed and not even breathing hard, without anyone finding it odd. The programs are likely to be better than expected during four months of the year: February, May, July, and November. Those are the months that Nielsen Media Research chooses as its "sweeps" months. Nielsen usually gauges ratings by electronically monitoring a small number of household TVs, but during these months, the company passes out a great number of viewing diaries. The diaries provide a detailed breakdown of viewers. If advertisers don't like the results, the broadcasters suffer. In an attempt to attract more viewers during the all-important months, local news programs are likely to broadcast their most popular features, while the national stations will show their best movies.

High-speed gushes of electromagnetic waves are spurting out from this TV set, as well as from the transmitting station. Some soak right into us, leaving behind about 1.5 millirems of radiation, which is the equivalent of three chest X rays each year—a figure that sometimes can be reduced by sitting slightly off-center from the screen, thereby making oneself less of a direct target. Most of the rippling electromagnetic wave fronts blast through the house walls, traveling immensely faster than the earlier escaping kitchen gases, and are soon hurtling upward to the stars. The result is a great bubble of escaped TV programs now expanding through interstellar space. Its outermost rim carries the grainy image of the first BBC broadcasts from Crystal Palace in north London, while inner bubbles, roaring away a mere 9,200 trillion miles from earth, spread the visages of Lucille Ball, Walter Cronkite, and Ozzie and Harriet.

Lost in the on-screen world, oblivious to the house and lawn beyond, the dad has no chance of noticing the discreet flying monster that is now hovering in the room. Female mosquitoes only come into our rooms to look around a little and fly next to our ears with a horrible irritating whine and then back off a little so we can't find the damned things, because they're tender, caring creatures who are willing to forego any danger to get high-quality iron-thickened water—our blood!—after their eggs have been fertilized. Male mosquitoes, happy-go-lucky wastrels, utterly without a sense of responsibility, never bother us this way. An occasional slurp of cheap, low-grade plant juice is all they need to keep going. Only female mosquitoes bite, because only females need the iron and amino acids and other nutrients that our blood vessels, unfortunately accessible to stab wounds from the surface, come so plentifully supplied with.

The mosquito at the doorway looks clumsy at first, but its body is well designed for tidily harvesting our blood. They're not even being clumsy when they wobble in place. Mosquitoes are so small that they don't have to actually bother with flying in the way we understand it. The air is thick enough at their microsize that it becomes as buoyant as a pond of water would be for us. All mosquitoes have to do is row their way across the air. What they're doing now is simply bobbing in position, tiny wings easily sculling, as they get their attack plan underway. Any male mosquito nearby would hear that sculling, as the whole air pond from our room stretching outward moves slightly in time with the wings. This is what allowed them to find the females for mating earlier. (Female mosquitoes that are too young for mating beat their wings more slowly, sending out slower ripples that the males ignore.)

The mosquito finds the dad by locating the carbon dioxide cloud he's emitting—like all nice blood-filled mammals, humans are crude combustion machines, constantly generating carbon dioxide as waste, which the mosquito can then aim toward. (Tormented caribou herds will hurry into forests, even when there may be wolves there, just to break up the carbon dioxide plumes that could otherwise signal dive-bombing mosquito flocks.) Even a slumped TV watcher will stir now, flapping one hand up, when he hears that distinctive mosquito whine approaching. Mosquitoes will often pause here, hanging in place a little, pumping tube only partly extended, to see what happens next. A second powerful air swish by the human though, more accurate this time, and now only flyers keen for personal investigation of the Mosquitan Afterlife will risk floating so close any longer: it flies away, to hunt for the blood it needs elsewhere.

Deeper into the couch the electronically bonding father sinks, and in his fascination, the physiological embarrassment reaches fresh depths. The basic metabolic slowdown takes over quickly enough, but if he's lugged his feet up onto the couch too, then his breathing slows even more. Gravity no longer tugs his abdomen downward, and the lungs take up a slight outward bulge, leaving a little more air trapped uselessly inside at each breath. The mouth slips open, so ending the effort of dragging air through the nose and pharynx, and when swallowing slows, the drooling that's characteristic of a daytime nap might begin.

It's about as close to being a member of the British royal family as one can get, and the innermost alveolar chambers of the lungs now begin to collapse. Babies that are very premature always have collapsed lungs, because the surface tension of liquids that are on the lung's inner surfaces tugs them